Existing Sewer Evaluation and Rehabilitation

WEF Manual of Practice FD-6

ASCE Manuals and Reports on Engineering Practice No. 62

Second Edition

Prepared by a Joint Task Force of the **Water Environment Federation** and the **American Society of Civil Engineers.**
B. Jay Schrock, *Chair*
Phyllis A. Brunner, *Vice-Chair*

Stephen E. Cordes	Graham Knott
Arun K. Deb	John Larson
Irene Dooley	David London
Roy C. Fedotoff	Boyd C. Mills
Rodolfo B. Fernandez	Kent Olson
John S. Gresh	Paul E. Simpson
Akhtar Hamid	Richard O. Thomasson
William A. Hunt	Harry N. Tuvel
Jey K. Jeyapalan	John Warburton
Kenneth K. Kienow	Stephen Waring

Under the Direction of the **WEF Facilities Development Subcommittee of the Technical Practice Committee** and **ASCE Environmental Engineering Division Committee on Water Pollution Management** and the **ASCE Pipeline Division Committee on Pipeline Infrastructure.**

1994

Water Environment Federation
601 Wythe Street
Alexandria, VA 22314-1994 USA

and

American Society of Civil Engineers
345 East 47th Street
New York, NY 10017-2398 USA

The material presented in this publication has been prepared in accordance with generally recognized engineering principles and practices and is for general information only. This information should not be used without first securing competent advice with respect to its suitability for any general or specific application.

The contents of this publication are not intended to be a standard of the Water Environment Federation (WEF) or the American Society of Civil Engineers (ASCE) and are not intended for use as a reference in purchasing specifications, contracts, regulations, statutes, or any other legal document.

No reference made in this publication to any specific method, product, process, or service constitutes or implies an endorsement, recommendation, or warranty thereof by WEF or ASCE.

WEF and ASCE make no representation or warranty of any kind, whether expressed or implied, concerning the accuracy, product, or process discussed in this publication and assume no liability.

Anyone using this information assumes all liability arising from such use, including but not limited to infringement of any patent or patents.

Library of Congress Cataloguing-in-Publication Data

Existing sewer evaluation and rehabilitation / prepared by a joint task force of the Water Environment Federation and the American Society of Civil Engineers; under the direction of the WEF Facilities Development Subcommittee of the Technical Practice Committee and ASCE Environmental Engineering Division Committee on Water Pollution Management.
    p. cm.—(ASCE manual and report on engineering practice; no. 62) (WEF manual of practice; FD–6)
    Includes bibliographical references and index.
    ISBN 1-881369-92-7: $50.00 — ISBN 0-7844-0049-0: $50.00
    1. Sewerage—Inspection. 2. Sewerage—Maintenance and repair.
I. Water Environment Federation. II. Amercian Society of Civil Engineers.
III. Water Environment Federation. Facilities Development Subcommittee.
IV. American Society of Civil Engineers. Committee on Water Pollution Management. V. Series. VI. Series: Manual of practice. FD; no. 6.
TD719.E95 1994   94-29428
628'.24—dc20   CIP

Copyright © 1994 by the Water Environment Federation and the American Society of Civil Engineers. Permission to copy must be obtained from both WEF and ASCE.

All Rights Reserved.

Library of Congress Catalog Card No. (94-29428)
ISBN (1-881369-92-7)
Printed in the USA **1994**

Water Environment Federation

The Water Environment Federation (WEF) is a nonprofit, educational organization composed of member and affiliated associations throughout the world. Since 1928, WEF has represented water quality specialists, including civil, design, and environmental engineers, biologists, bacteriologists, local and national government officials, treatment plant operators, laboratory technicians, chemists, industrial technologists, students, academics, and equipment manufacturers and distributors.

For information on membership, publications, and conferences contact
Water Environment Federation
601 Wythe Street
Alexandria, VA 22314–1994 USA
(703) 684-2400

American Society of Civil Engineers

The American Society of Civil Engineers (ASCE) offers civil engineering professionals many opportunities for technical advancement, networking, and leadership and technical skill training. Also available to members are major savings on educational seminars, conferences, conventions, and publications. First class, low-cost insurance programs are among the most competitive available.

Members may participate on a national level, networking with colleagues on forums to advance the profession. This participation affords the opportunity to develop leadership skills and to expand personal contacts. On the local level, chapters (called Sections and Branches) act as advocates in the public interest on local issues and present seminars and programs relevant to the needs of the local community. For more information call 800-548-ASCE.

Manuals of Practice for Water Pollution Control

(*As developed by the Water Environment Federation*)

The Water Environment Federation (WEF) Technical Practice Committee (formerly the Committee on Sewage and Industrial Wastes Practice of the Federation of Sewage and Industrial Wastes Associations) was created by the Federation Board of Control on October 11, 1941. The primary function of the committee is to originate and produce, through appropriate subcommittees, special publications dealing with technical aspects of the broad interests of the Federation. These manuals are intended to provide background information through a review of technical practices and detailed procedures that research and experience have shown to be functional and practical.

Water Environment Federation Technical Practice Committee Control Group

F.D. Munsey, *Chair*
L.J. Glueckstein, *Vice-Chair*

C.N. Lowery
T.L. Krause
T. Popowchak
J. Semon

Authorized for Publication by the Board of Control
Water Environment Federation

Quincalee Brown, *Executive Director*

ASCE Manuals and Reports on Engineering Practice

(*As developed by the ASCE Technical Procedures Committee July 1930 and revised March 1935, February 1962, and April 1982*)

A manual or report in this series consists of an orderly presentation of facts on a particular subject, supplemented by an analysis of limitations and applications of these facts. It contains information useful to the average engineer in his everyday work, rather than the findings that may be useful only occasionally or rarely. It is not in any sense a "standard," however, nor is it so elementary or so conclusive as to provide a "rule of thumb" for nonengineers.

Furthermore, material in this series, in distinction from a paper (which expresses only one person's observations or opinions), is the work of a committee or group selected to assemble and express information on a specific topic. As often as is practicable, the committee is under the general direction of one or more of the Technical Divisions and Councils, and the product evolved has been subjected to review by the Executive Committee of that Division or Council. As a step in the process of this review, proposed manuscripts are often brought before the members of the Technical Divisions and Councils for comment, which may serve as the basis for improvement. When published, each work shows the names of the committee by which it was compiled and indicates clearly the several processes through which it has passed in review so its merit may be definitely understood.

In February 1962 (and revised in April 1982), the Board of Direction voted to establish

> A series entitled "Manuals and Reports on Engineering Practice" to include the Manuals published and authorized to date, future Manuals of Professional Practice, and Reports on Engineering Practice. All such Manual or Report material of the Society would have been refereed in a manner approved by the Board Committee on Publications and would be bound, with applicable discussion, in books similar to past Manuals. Numbering would be consecutive and would be a continuation of present Manual numbers. In some cases of reports on joint committees, bypassing of Journal publications may be authorized.

American Society of Civil Engineers

H. David Stensel, *Chair,* Environmental Engineering Division, 1994
J.P. Castronovo, *Chair,* Pipeline Division, 1994
Paul T. Bowen, *Chair,* Water Pollution Management Committee
H. David Stensel, *Chair,* Manual Review Panel

Authorized for Publication by the Publications Committee of the American Society of Civil Engineers

Edward O. Pfrang, *Executive Director,* ASCE
James E. Davis, *Executive Director Designate*, ASCE

Abstract

This manual, *Existing Sewer Evaluation and Rehabilitation, Second Edition*, provides general guidance and serves as a source of information in the evaluation and rehabilitation of existing sewers. Recent technological advances, such as high-resolution closed-circuit television and improved analytical methods and materials make rehabilitation of existing sewers a cost-effective alternative to relief sewer construction. With these advances in mind, this manual explores three major topics: sewer evaluation and monitoring, sewer rehabilitation methods and materials, and quality assurance. After a short explanation of the purpose and scope of the manual, Chapters 2, 3, 4, and 5 discuss techniques used for evaluation of existing sewers. The techniques covered are the general approach, methods of structural evaluation, and methods of infiltration and inflow evaluation. In addition, the effects of combined sewer overflow on pollution are described. Chapter 6 discusses sewer flow, quality monitoring, and sewer system hydraulics. Methods and materials used for pipeline rehabilitation are presented in Chapters 7 and 8. The final chapter discusses the quality assurance of long-term performance in sewer rehabilitation. A general formulae section and glossary are appended.

Preface

This manual of practice is intended to be a comprehensive and useful source of existing sewer evaluation and rehabilitation information for the practicing design engineer. The manual replaces *Existing Sewer Evaluation and Rehabilitation* published jointly in 1983 by the Water Environment Federation (formerly the Water Pollution Control Federation) and the American Society of Civil Engineers.

The major emphasis of this manual is on infiltration and inflow reduction, with lesser emphasis placed on maintaining the structural integrity of the existing sewer. Many of the rehabilitation methods presented are applicable for both aspects of sewer rehabilitation. The structural integrity aspects of sewer rehabilitation cannot be overemphasized because of the potentially destructive and costly events that can result after a sewer system collapse.

The contents of this manual represent the collected background and experience of professionals active in the evaluation and rehabilitation of existing sewers.

This manual was produced under the direction of B. Jay Schrock, *Chair*. The principal authors are

Phyllis A. Brunner
Rodolfo B. Fernandez
John S. Gresh
Graham Knott

B. Jay Schrock
Harvey Swetlik
John Warburton

American Society of Civil Engineers review was performed by Wayne Dillard, H. David Stensel, and Cecil Lue-Hing.

Authors' and reviewers' efforts were supported by the following organizations:

American Ventures, Inc.
Black & Veatch
Boswell Engineering
Brown & Caldwell
Central Contra Costa Sanitary District
CH2M Hill
Clark County Sanitation District
Gelco
HKM Associates
JSC International Engineering
Kienow Associates

Metro St. Louis Sewer District
Montgomery Watson
RJN Group, Inc.
Stearin, Preston & Burrows
Sverdrup Corporation
U.S. Environmental Protection Agency
Washington Suburban Sanitary Commission
Weston Environmental Consultants
Woodward-Clyde Consultants

Federation technical staff project management was provided by Greg White and Maureen Ross; technical editorial assistance was provided by Bill Wolle.

Contents

Chapter		Page
1	Introduction	1
	Overview and Historical Background	1
	Problems and Management	2
	Purpose and Scope	2
	The Need for Guidelines	3
	Format	4
	Reference	4
2	General Approach	5
	Types of Systems	6
	Sanitary	6
	Storm	6
	Combined	6
	Problems	7
	Structural	8
	Collapse Mechanisms	9
	Hydraulic	9
	Infiltration	9
	Inflow	12
	Combined Sewer Overflow Effects	12
	Flooding	13
	Environmental	13
	Critical Sewers	13
	Categories	13
	Critical Definitions	14
	Performance Criteria	15
	Integrated Approach	17
	Need	17
	Planning Investigations (Phase 1)	17
	Review System Records	17
	Categorize Sewers	17
	Document Performance	17
	Select a Detailed Investigation and Establish Priorities	18

	Improve System Records and Access	18
	Assessing the System Condition (Phase 2)	19
	Assessing the Infiltration and Inflow Condition (Phase 2-A)	20
	Infiltration and Inflow Condition	20
	Flow Monitoring	20
	Qualify and Quantify Extraneous Flows	20
	Assessing the Structural Condition (Phase 2-B)	20
	Structural Condition	20
	Evaluate Corrosion	20
	Qualify and Quantify Problem Areas	21
	Assessing Hydraulics (Phase 2-C)	21
	Hydraulics	21
	Build Hydraulic Model	21
	Confirm Field Conditions	21
	Calibrate Model	21
	Assess Hydraulic Performance	21
	Locate Areas of Performance Deficiencies	21
	Assembling Data	21
	Identifying Systems Needing Rehabilitation or Replacement	21
	Developing the System Usage Plan (Phase 3)	22
	Setting Priorities for Each Problem	22
	Considering Rehabilitation Options or Replacement	22
	Developing Consistent Solutions to Problems	22
	Identifying Cost-Effective Solutions	22
	Establishing the System Usage Plan	22
	Implementing the System Usage Plan (Phase 4)	22
	Designing and Constructing Rehabilitation and Replacement Projects	22
	Monitoring Conditions of Primary Sewers	23
	Adjusting the Hydraulic Model as Needed	23
	Reviewing Usage Plan as Needed	23
	Outline of Cursory Investigation	23
	Predictive Modeling	23
	Necessary Functions	23
	Hydrologic Function	23
	Hydraulic Function	25
	Modeling Time Scale	25
	Justification for Modeling	26
	References	26
3	Methods of Structural Evaluation	27
	Pipeline Conditions	27

	Structural Conditions	28
	Corrosion Conditions	28
	Qualification and Quantification	28
	Brick Sewers	28
	Concrete and Clay Sewers	29
Rehabilitation Assessment		30
Program Plan		32
Evaluation Assessment		32
	Monitoring	33
	Stabilization	34
	Rehabilitation	34
	Replacement	35
References		35

4 Methods of Infiltration and Inflow Evaluation 37

- Purpose 38
- Problems 40
- Evaluation Techniques 40
  - Flow Components 40
    - Peak Rate Determinations 41
    - Annual Volume Determination 41
    - Preinstallation Considerations 42
    - Meter Maintenance 43
    - Data Evaluation 44
  - Precipitation Measurement 44
  - Groundwater Gauging 45
  - Smoke Testing 46
    - Procedure 47
    - Photographs 49
  - Manhole and Pipeline Visual Inspection 49
    - Inspection 49
    - Safety Measures 50
      - Atmospheric Hazards 51
        - Explosive Gases 51
        - Toxic Conditions 51
        - Oxygen Deficiency 51
      - Physical Injury 51
      - Infections 52
      - Animals 52
      - Chemicals 52

Drowning	52
Data Recording	53
Building Inspection	53
Building Inspection Program Procedure	54
Dye-Water Testing	55
Determining Conditions Caused by Storm Sewer Sections	56
Determining Conditions Caused by Stream Sections	56
Identifying Sources on Private Property	56
Identifying Structurally Damaged Manholes	56
Verifying Sources Found by Other Means	57
Safety Measures	57
Logging Test Results	58
Night Flow Isolation	58
Groundwater Migration	59
Single-Pass Method	60
Multipass Method	60
Isolation Techniques	60
Plugging	60
Differential Isolation	62
Portable Weir Methods	62
Velocity-Area Method	63
Fluorometric Methods	64
Results	64
Pipeline Cleaning and Television Inspection	65
Photographs	66
Leak Quantification	66
Evaluation of Infiltration	68
Water Use Evaluation	68
Maximum–Minimum Daily Flow	69
Maximum Daily Flow Comparison	69
Nighttime Domestic Flow Evaluation	70
Small Subsystems	70
Infiltration Location	71
Groundwater Levels	72
Interviews	72
Visual Inspection	72
Cost Effectiveness	72
Evaluation of Inflow	73
Applicable Techniques	74
Methodology	74
Rainfall-Induced Inflow Localization	77
Cost Effectiveness	78

	Evaluation of Physical Conditions	79
	Structural Integrity	79
	Operational and Maintenance Problems	80
	Map Correction and Updating	80
	Applicable Techniques	81
	Visual Inspection	81
	Television Inspection	82
	References	83
5	**Combined Sewer Overflow Pollution Abatement**	85
	History and Problems	86
	Regulatory Perspective	86
	Planning	87
	Issues	87
	Environmental Setting	87
	Regulatory Climate	88
	Infrastructure	88
	Sociopolitical Setting	88
	Data Requirements	89
	Analytical Tools	89
	Investigation	89
	Mapping	90
	Field Investigations	90
	Flow Monitoring	90
	Sampling	91
	Modeling	91
	Quantity and Quality	92
	Model Building and Verification	92
	Assessing Performance	93
	Determining the Pollutant Load	94
	Identifying the Location and Causes of Deficiencies	94
	Upgrading and Treatment Options	94
	Source Control	94
	Sewer System Control	95
	Attenuating Peak Flows	95
	Increasing System Capacity	95
	Real-Time Control	96
	Treatment Strategies	96
	Screening	96
	Sedimentation	97

	Chemical Treatment	97
	Filtration	97
	Air Flotation	97
	Disinfection	97
References		98
6	Flow Monitoring	99
	Purpose	100
	General Data Needs	100
	Maps of Study Area	100
	Records	101
	Sewer Maps	101
	Sewer Conditions	101
	Flow Records	101
	Bypass and Overflow Information	101
	Emergency Pumping	101
	Water Usage	102
	Rainfall Data	102
	Groundwater, Lake, and Stream Data	102
	Demography	102
	Recordkeeping Program	102
	Monitoring Program	103
	Research	103
	Site Selection	104
	Key Manholes	104
	Bypasses	105
	Overflows	105
	Factors Affecting Flow	105
	Equipment Selection and Sizing	106
	Timing and Data Correlation	106
	Safety Program	107
	Monitoring Duration	107
	Instantaneous Monitoring	108
	Random Monitoring Interval	108
	Continuous Short-Term Monitoring	108
	Continuous Long-Term or Permanent Monitoring	109
	Available Equipment	109
	Measurement Theory	109
	Manning's Equation	110
	Calibrated Discharge Curves (Stage Discharge Curves)	110
	Velocity	110

Manual Measurements	111
Weirs	111
Triangular (V-Notch) Weir	111
Rectangular (Contracted) Weir With End Contractions	112
Rectangular (Suppressed) Weir Without End Contractions	112
Trapezoidal (Cipolletti) Weir	113
Compound Weir	113
Flumes	114
Parshall Flume	115
Palmer–Bowlus Flume	116
H-Flume	117
Trapezoidal Flume	117
Dye, Chemical, and Radioactive Tracers	117
Pumping Station Calibration	118
Calibration Using Wet Well Drawdown or Return	118
Calibration Using a Velocity Meter	119
Calibration Using Discharge Volume	119
Calibration Using a Pump Curve	120
Recording Calibration	120
Bucket Test	121
Stage Measurement	121
Automatic Flow Meters	121
Depth Recorders	121
Probe Recorders	121
Bubbler Recorders	122
Pressure Sensors	122
Float Recorders	122
Ultrasonic Recorders	122
Capacitance or Electronic Recorders	122
Velocity Meters	123
Doppler Meters	123
Orifice, Nozzle, and Venturi Meters	123
Current Meters	123
Velocity Probes	124
Site Work	124
Maintaining Metering Equipment	124
Manual Measurements	124
Continuously Recording Measurements	126
Depth-Velocity Meters	126
Depth-Only Meters	126
Measuring Groundwater	128
Measuring Rainfall	129

Data Analysis	129
Editing Data	129
Filtering	129
Offset	129
Time Adjustment	129
Viewing Data	129
Determining Key Values	129
Quality Assurance	130
Manual Verification of Meter Data	130
Review of Recorded Data	130
Problem Areas	130
References	131
7 Pipeline Rehabilitation Methods	**133**
General Considerations	135
Functional Overlap	135
Durability or Life Expectancy	141
Long-Term Material Properties	144
Resistance to Chemical Attack	144
Abrasion Resistance	145
Structural Considerations	145
Loadings	145
Design Types	145
Hydraulic Capacity	145
Cost	147
Systemwide Implications	147
Construction Issues	148
Safety	148
Confined Spaces	148
Trenches	148
Adjacent Structures and Utility Plants	148
Preparation	148
Flows	149
Sewer Cleaning	149
Voids or Loose Ground	149
Methods of Working	149
Materials	149
Pipe Laying	150
Annular Grouting	150
Supervision	150
Contract Documentation	151
Pipeline Renovation Systems	152

Pipe Linings	152
Sliplining	152
Continuous Pipe	152
Short Pipes	153
Solid-Wall Polyethylene	154
Profile Wall Polyethylene	154
Spiral Rib Polyethylene	155
Polyvinyl Chloride	155
Fiber-Glass-Reinforced Plastic	155
Reinforced Plastic Mortar	156
Cement-Lined or Polyethylene-Lined Ductile Iron	156
Steel	156
Cured-in-Place Pipe	156
Nu-Pipe and U-Liner Deformed Pipe	158
Rolldown and Swagedown Deformed Pipe	159
Spiral-Wound Pipe	161
Segmental Linings	161
Fiber-Glass-Reinforced Cement	161
Fiber-Glass-Reinforced Plastic	162
Reinforced Plastic Mortar	162
Polyethylene	162
Polyvinyl Chloride	162
Steel	162
Structural and Nonstructural Coatings	162
Reinforced Shotcrete	163
Cast-in-Place Concrete	163
Pipeline Replacement or Renewal	164
Trenchless Replacement	165
Pipe Bursting	165
Advantages	167
Disadvantages	168
Microtunneling	169
Auger Systems	169
Slurry Systems	170
Pipe Installation	170
Advantages	171
Disadvantages	172
Other Trenchless Systems	172
Directional Drilling	172
Fluid Jet Cutting	172
Impact Moling	172
Impact Ramming	172

Auger Boring	173
Conventional Replacement	173
Maintenance and Repair	**175**
Cleaning	175
Jet Rodding	176
Rodding	176
Winching or Dragging	176
Cutting	176
Manual or Mechanical Digging	176
Root Control and Removal	176
Pointing	178
Internal Grouting	178
Small- and Medium-Sized Pipes	178
Large-Diameter Pipe Grouting	179
Materials	181
Advantages	181
Limitations	182
Cost and Feasibility	182
External Grouting	183
Chemical Grouts	183
Cement Grout	184
Portland Cement Grout	184
Microfine Cement Grout	185
Compaction Grouting	185
Mechanical Sealing	185
Point (Spot) Repairs	185
Manhole Rehabilitation	**187**
Manhole Conditions	187
Structural Degradation	187
Movement and Displacement	187
Corrosive Environments	188
Excessive Infiltration and Inflow	188
Maintenance	188
Manhole Rehabilitation Methods	188
Chemical Grouting	189
Coating Systems	189
Structural Linings	191
Corrosion Protection	193
Frame, Cover, and Chimney Rehabilitation	193
Service Connection Rehabilitation	**194**
Rehabilitation Methods	195
Chemical Grouting	196

	Cured-In-Place Pipe Lining	196
	Other Measures	197
	References	197
8	**Pipeline Rehabilitation Materials**	**199**
	Sliplining	200
	Continuous Pipe	200
	High-Density Polyethylene Pipe	200
	Polybutylene Pipe	200
	Short Pipe	201
	High-Density Polyethylene Pipe	201
	Polyvinyl Chloride Pipe	201
	Fiber Glass Pipe	202
	Ductile Iron Pipe	203
	Steel Pipe	203
	Cured-In-Place Pipe	203
	Deformed Pipe	204
	U-Liner Pipe	204
	Nu-Pipe	204
	Rolldown	204
	Spiral-Wound Pipe	205
	Segmental Linings	206
	Fiber-Glass-Reinforced Cement	206
	Fiber-Glass-Reinforced Plastic	206
	Reinforced Plastic Mortar	206
	Polyethylene	206
	Polyvinyl Chloride	206
	Welded Steel	207
	Coatings	207
	Replacement Materials	207
	Maintenance and Repair Materials	207
	Root Control	207
	Chemical Grouting	207
	Acrylamide Grout	208
	Acrylic Grout	209
	Acrylate Grout	209
	Urethane Grout	209
	Urethane Foam	210

	Reference	210
9	Quality Assurance	211
	Setting the Objectives of Sewer Rehabilitation	213
	Infiltration and Inflow Control Objectives	213
	Selecting the Design Condition	213
	Allocating System Components for Quantified Infiltration and Inflow	213
	Estimating the Effectiveness of Rehabilitation for Infiltration and Inflow Reduction	214
	Structural Rehabilitation Objectives	215
	Construction-Phase Quality Assurance	215
	Material Quality Control	215
	Existing Sewer Conditions	216
	Safety	216
	Preparation of Sewers	216
	Installation	216
	Annular Grouting	216
	Contractor Payment	217
	Construction Inspection	217
	Measuring the Effectiveness of Infiltration and Inflow Control	217
	Methods for Normalizing Flow Data	218
	Infiltration Simulation	218
	Inflow Simulation	219
	Sanitary Flow Simulation	219
	Storm Inflow Comparison	219
	Flow-Monitoring Considerations	220
	Studies of Rehabilitation Effectiveness for Infiltration and Inflow Control	220
	Measuring the Effectiveness of Rehabilitation for Structural Integrity	221
	Specific Repair Effectiveness	221
	Overall System Integrity	221
	Expected Effectiveness of Special Sewer Rehabilitation Methods	221
	Sewer Replacement	221
	Sewer Relining	222
	Lining and Sliplining	222
	Inversion Lining	223
	Sewer Sealing	223
	Service Lateral Rehabilitation	224
	Inflow Control	224

Manholes	224
Catch Basins	225
Roof Drains	225
Other	225
Overall Sewer System Effectiveness of Infiltration and Inflow Control	225
Continuing Sewer Maintenance	226
Data Needs for Budgeting Preventive Maintenance	227
Sewer Maintenance Activities	228
Annual Reporting	228
References	228
Appendix A General Formulae	231
Appendix B Glossary	237
Index	263

List of Tables

Table		Page
2.1	Sewer system condition measures.	7
3.1	Brick sewer structural condition evaluation criteria.	29
3.2	Concrete and clay sewer structural condition evaluation criteria.	31
3.3	Internal condition rating factors.	32
7.1	Pipeline rehabilitation options—renovation.	136
7.2	Pipeline rehabilitation options—minimum excavation and trenchless replacement.	138
7.3	Pipeline rehabilitation options—conventional replacement and maintenance or repair.	139
7.4	Manhole rehabilitation options.	142
7.5	Service connection rehabilitation options.	143
7.6	General comparison of durability of pipeline.	144
7.7	Typical shotcrete and gunite mixes.	165
7.8	Pipe bursting replacement method.	167
7.9	Ranking of manhole infiltration sources.	188
7.10	Miscellaneous private property rehabilitation measures.	197
8.1	Primary properties of concern for polyethylene sliplining pipe.	201
8.2	Primary properties of concern for polyvinyl chloride sliplining pipe.	202
8.3	Primary properties of concern for cured-in-place pipe.	204
8.4	Primary properties of concern for polyvinyl chloride deformed pipe.	205
8.5	Dimensions of polyvinyl chloride strips and panels.	205
9.1	Example collection system component infiltration allocation.	214

List of Figures

Figure		Page
2.1	Proper pipe side support.	8
2.2	Deformation of cracked pipes.	10
2.3	Subsidence of sewer.	11
2.4	Collection system components and infiltration and inflow sources.	12
2.5	Critical sewer identification process.	15
2.6	Detailed sewer investigation.	18
2.7	Cursory investigation.	24
4.1	Typical weekly hydrograph for a monitoring site, South St. Paul, Minnesota.	45
4.2	Static groundwater manhole gauge installation elevation.	46
4.3	Sample smoke testing notice.	48
4.4	Daily component flow comparison.	70
4.5	Hydrographs—the most common method of presenting flow data.	76
4.6	Cost curve for determination of inflow sources to be rehabilitated.	79
6.1	Flow-monitoring weirs.	112
6.2	Weir details and installation.	113
6.3	Parshall flume.	115
6.4	Palmer–Bowlus flume details.	116
6.5	Manual weir installation.	125
6.6	Temporary continuous flow recorder installation.	127
7.1	Traditional sliplining.	153
7.2	Welding in trench.	153
7.3	Insertion method for sliplining with short pipes.	154
7.4	Cured-in-place pipe installation.	157

7.5	Deformed pipe.	159
7.6	General arrangement of rolldown and insertion.	160
7.7	Swagelining.	160
7.8	Gunite or shotcrete wall construction.	164
7.9	Pipe bursting—typical site layout.	166
7.10	Microtunneling (auger system).	170
7.11	Microtunneling (slurry system).	171
7.12	Directional drilling.	173
7.13	Schematic arrangement of fluid jet cutting system.	174
7.14	Internal grouting equipment.	179
7.15	Internal grouting of large-diameter pipe joints.	180
7.16	Mechanical seal.	186
7.17	Typical section through rehabilitated manhole.	192
7.18	Polyethylene manhole insert.	194

Chapter 1
Introduction

1 Overview and Historical Background
2 Problems and Management
2 Purpose and Scope
3 The Need for Guidelines
4 Format
4 Reference

OVERVIEW AND HISTORICAL BACKGROUND

The primary purpose of sewer rehabilitation originally was to maintain the structural integrity of a sewer system for dependable transfer of wastewater from the source to the treatment plant or receiving water. If this were the only purpose, concern about effectiveness of wastewater transfer would be minimal, other than for structural reasons.

With the passage of the Federal Water Pollution Control Act Amendments of 1972, however, more emphasis was placed on reducing the hydraulic loads on treatment plants caused by excessive infiltration and inflow (I/I). Reduced I/I can significantly reduce hydraulic loading into collection and treatment facilities during periods of wet weather. This lowers capital costs associated with oversized facilities, reduces operation and maintenance costs, and prolongs the life of treatment and conveyance facilities.

The U.S. Environmental Protection Agency's (U.S. EPA) 1972 requirement for I/I analysis when preparing facility plans indicates the emphasis being placed on sewer rehabilitation. This manual is a guide for the person or agency confronted with evaluating and rehabilitating existing sewer facilities.

Recent technological advances, such as high-resolution, closed-circuit television cameras, have made it possible to observe the condition of existing sewers. Combined with improved analytical methods and materials, this technology has made rehabilitation of existing sewers a cost-effective alternative to relief sewer construction or replacement.

PROBLEMS AND MANAGEMENT

Leaking underground sanitary sewer systems can cause exfiltration of raw wastewater and industrial discharge through leaking pipes, contaminating soil and groundwater. In other cases, infiltration of groundwater through leaky joints and cracks in the pipe system can lead to excess cost at the treatment plant or contribute to pipe collapse.

Billions of dollars have been spent at all levels of government in the U.S. to meet the requirements of the Clean Water Act, though few of U.S. EPA's construction grants have gone toward the rehabilitation or replacement of the 19 800 collection systems around the country (U.S. EPA, 1972). In the U.S., there remain approximately 1 100 combined sewer systems and 10 000 to 15 000 combined sewer overflow (CSO) discharge points. Funding for these needs is limited, and a deferred maintenance, out-of-sight, out-of-mind philosophy still prevails in many regions.

Yet ever-improving technologies are now in place that can provide cost-effective solutions to underground infrastructure problems. The following steps might help preserve underground sewer systems in the U.S.:

- Assess the condition of the current piping system, evaluate proven cost-effective alternatives for repairs, and budget accordingly. Develop a master plan for phased implementation beginning with the most critical portion of the system.
- Examine user fees and develop programs for matching them to assessed needs.
- Evaluate alternative funding, including grants, loans, and creative financing techniques.
- Develop a creative communications program that will carry the message to the voting public and others who can influence funding appropriations and rates.
- Institute programs on a timely basis, adjust as required, and report progress to the public, pointing out savings received and other benefits.

PURPOSE AND SCOPE

The purpose of this manual is to provide general guidance and a source of information for the evaluation and rehabilitation of existing sewers. Accordingly, typical procedures, case studies, and examples are presented. The information is intended for use by engineers, municipalities, regulatory agencies, and others with responsibility for planning, designing, constructing, financing, regulating, operating, or maintaining sewer systems.

This manual provides information on conducting I/I analyses, sewer system evaluation surveys, and rehabilitation methodology. It does not contain standards, rules, or regulations. Rather, the design information presented is intended as general guidance reflecting sound, professional practice.

The technologies discussed in this manual were selected through past experience and the availability of information and performance data. The use of any method or material does not guarantee performance, and the omission of a particular method does not reflect on its acceptability. All viable techniques should be considered when evaluating and planning the reconditioning of existing sewers.

This manual discusses

- Existing sewer system problems;
- Techniques for evaluating the performance of existing sewers;
- Methods for evaluating information gathered in testing;
- Techniques for measuring flow;
- The design of a system-monitoring program;
- A procedure for selecting the method for an overall approach to sewer rehabilitation;
- Advantages, limitations, disadvantages, costs, and feasibility of pipeline rehabilitation;
- Various methods of pipeline reconditioning;
- Rehabilitation for manholes and service laterals;
- Materials used for rehabilitating pipelines, manholes, and service laterals; and
- Quality assurance.

THE NEED FOR GUIDELINES

As promulgated by the Federal Water Pollution Control Amendments of 1972, regulatory guidelines to control sewer system effluent via the study, operation, maintenance, and repair of sewer systems indicated that available technical guidelines were inadequate.

The success of sewer system rehabilitation in reducing I/I as evaluated by U.S. EPA concluded that the correlation of the diverse methods of prerehabilitation study and postrehabilitation evaluation was impossible. The need for state-of-the-art technical guidelines for use in the development of comprehensive sewer rehabilitation programs became evident.

Engineering practice in response to regulatory guidelines has matured over the past 20 years. This manual is intended to improve sewer rehabilitation success and provide a measure against which future rehabilitation methods may be compared.

Format

In this manual, effort has been made to address, chronologically, the development of analyzing the I/I, assessing structure, evaluating sewer system surveys, and outlining the methods available for rehabilitation and their use.

Chapters 2, 3, 4, and 5 discuss evaluation techniques: the general approach, evaluation of structural data, evaluation and location of infiltration data, and evaluation of rainfall-induced infiltration, inflow, and CSO impacts. Chapter 6 discusses sewer flow and quality monitoring and provides information needed to determine the hydraulics in the sewer system. Chapters 7 and 8 describe rehabilitation techniques and materials currently used for sewer systems. Various considerations are provided for the selection of a particular rehabilitation method for pipelines, manholes, and service laterals. Chapter 9 discusses the objective of sewer rehabilitation and quality assurance of long-term performance.

A general formulae section is included for designing various performance parameters, and the glossary includes terms used frequently throughout the manual.

Reference

U.S. EPA (1972) Public Law 92-500, Washington, D.C.

Chapter 2
General Approach

6	Types of Systems	20	Assessing the Structural Condition (Phase 2-B)
6	Sanitary		
6	Storm	20	Structural Condition
6	Combined	20	Evaluate Corrosion
7	Problems	21	Qualify and Quantify Problem Areas
8	Structural		
9	Collapse Mechanisms	21	Assessing Hydraulics (Phase 2-C)
9	Hydraulic	21	Hydraulics
9	Infiltration	21	Build Hydraulic Model
12	Inflow	21	Confirm Field Conditions
12	Combined Sewer Overflow Effects	21	Calibrate Model
13	Flooding	21	Assess Hydraulic Performance
13	Environmental	21	Locate Areas of Performance Deficiencies
13	Critical Sewers		
13	Categories	21	Assembling Data
14	Critical Definitions	21	Identifying Systems Needing Rehabilitation or Replacement
15	Performance Criteria		
17	Integrated Approach		
17	Need	22	Developing the System Usage Plan (Phase 3)
17	Planning Investigations (Phase 1)		
17	Review System Records	22	Setting Priorities for Each Problem
17	Categorize Sewers	22	Considering Rehabilitation Options or Replacement
17	Document Performance		
18	Select a Detailed Investigation and Establish Priorities	22	Developing Consistent Solutions to Problems
18	Improve System Records and Access	22	Identifying Cost-Effective Solutions
19	Assessing the System Condition (Phase 2)	22	Establishing the System Usage Plan
		22	Implementing the System Usage Plan (Phase 4)
20	Assessing the Infiltration and Inflow Condition (Phase 2-A)		
		22	Designing and Constructing Rehabilitation and Replacement Projects
20	Infiltration and Inflow Condition		
20	Flow Monitoring		
20	Qualify and Quantify Extraneous Flows	23	Monitoring Conditions of Primary Sewers

23	Adjusting the Hydraulic Model as Needed	23	Hydrologic Function
23	Reviewing Usage Plan as Needed	25	Hydraulic Function
23	Outline of Cursory Investigation	25	Modeling Time Scale
23	Predictive Modeling	26	Justification for Modeling
23	Necessary Functions	26	References

Wastes originating from domestic, commercial, and industrial sources (often mixed with stormwater) are collected, treated, and discharged back to the environment. Protecting wastewater collection and treatment systems with the least risk to public health and safety in the most cost-effective manner is the goal of any sewer rehabilitation program.

Design, construction, maintenance, and rehabilitation of sewer infrastructures are important challenges. Because of their low visibility, underground utilities frequently get neglected until they suffer a catastrophic failure, which may be inconvenient and costly to repair. Periodic maintenance would be more cost effective in protecting public health.

This chapter describes types of sewer systems and their potential structural and hydraulic problems. Also, it suggests a critical sewer-monitoring approach, discusses performance criteria, summarizes an integrated rehabilitation strategy, and reviews predictive modeling tools used for evaluating sewer systems.

*T*YPES OF SYSTEMS

Sewers are conduits that carry wastewater or drainage water from an area. Sewer collection systems consist of sanitary, storm, or combined conduits.

SANITARY. A sanitary sewer carries waterborne wastes containing minor quantities of inadvertent storm, surface, and groundwater from residences, commercial buildings, industrial plants, and institutions.

STORM. A storm sewer carries storm runoff, along with street waste and wash water or drainage. It excludes domestic and industrial wastewater.

COMBINED. Typically found in older cities, combined sewers are collection systems that carry a mixture of domestic and industrial wastewater along with storm runoff.

During storms, when combined waste and runoff water exceed the capacity of a conveyance system, overflows are bypassed directly to the receiving watercourse without treatment. These overflows result in significant pollution and create health hazards.

PROBLEMS

Severe or catastrophic sewer collapses are rare, though are becoming more frequent. Disruptions, adverse publicity and public apprehension about potential recurrence normally follow sewer collapses. This type of failure is often associated with difficult ground conditions, large wastewater flows, adjacent utility impacts, traffic congestion, and deep excavation. Subsequent investigations often reveal incomplete records, infrequent inspections, or a failure to remedy known defects as factors contributing to the failure.

The Urban Institute (1981) conducted a study for the U.S. Department of Housing and Urban Development concerning sewer pipeline stoppages and collapses, which is summarized in Table 2.1.

This summary of 20 cities revealed a 1 000-to-1 spread on major main breaks and a 150-to-1 spread on stoppages per 1 600 km (1 000 miles) of sewers per year. Age and neglect are the most important reasons for these differences.

Table 2.1 Sewer system condition measures.

City	Annual major main breaks per 1 000 miles[a]	Annual stoppages per 1 000 miles[a]
Kansas City	219	325
Cleveland	97	—
Tulsa	93	363
Buffalo	77	505
Baltimore	73	296
Hartford	62	102
Atlanta	59	463
Charlotte	54	315
St. Louis	50	125
Des Moines	46	1 017
San Jose	45	1 265
Cincinnati	37	315
Philadelphia	31	—
Newark	25	—
Minneapolis	20	811
New York	16	3 629
Detroit	14	49
Denver	1.5	57
Seattle	0.4	25
Milwaukee	0.2	24
Mean	53.5	—
Median	48.1	—

[a] mile × 1.609 = km.

Roots that either puncture pipes or grow inside them cause more than 50% of the stoppages. Most structural failures are caused by roots, corrosion, soil movement, and inadequate construction combined.

There are an estimated 965 000 km (600 000 miles) of sanitary and combined sewers in the U.S. and approximately 50 major main breaks per 1 600 km/a (1 000 miles/year) for a total of 30 000 breaks annually.

According to the 1981 Urban Institute study, approximately 500 stoppages occur per 1 600 km (1 000 miles) per year, amounting to an estimated 300 000 stoppages annually. The study indicated that an additional 100 000 stoppages yearly were the responsibility of private residents.

The Water Research Centre has researched trends in the U.K. during the past 10 years and has indicated a 3% annual increase in problems (*Sewerage Rehabilitation Manual*, 1986). Approximately 5 000 collapses and 200 000 blockages occur each year. Collapse repair costs are dominated by the most expensive incidents, and repair costs are greater than those for new construction. Major breaks and stoppages occur more frequently in older sewer systems, many of which have had little, if any, preventive maintenance (Institution of Civil Engineers, 1981).

STRUCTURAL. In the past, sewers were constructed mainly of vitrified clay, brick, and concrete. Modern sewers include plastic, ductile iron, steel, and reinforced concrete. These materials typically have adequate compressive strength, and some have tensile strength. Rigid pipe materials are usually designed to resist vertical loading on their own, while brick sewers and flexible materials require side support from the surrounding soil (Figure 2.1).

Understanding the interaction between a sewer and the surrounding soil is important when assessing structural condition. The stability of a deteriorated

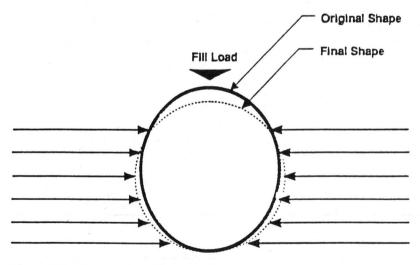

Figure 2.1 Proper pipe side support.

sewer depends on the lateral support provided by adjacent soil conditions. Depending on the size of the crack and the adjacent soil type, soil particles may pass through sewer defects.

Collapse Mechanisms. Factors that can contribute to deterioration and lead to structural failure include the size of the defect, soil type, interior hydraulic regime, groundwater level and fluctuation, corrosion, method of construction, and loading on a sewer. A collapse normally stems from deterioration at the site of an initial defect.

Initial defects usually are the result of poor construction, excessive loading, leaky joints, inadequate connections, or third party interference. When a defect is present, deterioration involving the adjacent soil takes place. If outside water breaches the sewer wall, soil can enter with it, resulting in ground loss and reduced side support. Collapse may be imminent and can be triggered by a random event adjacent to the area. Predicting the time of collapse is difficult because typically it is linked to an adjacent excavation or storm event and associated with a multitude of variables.

Collapse scenarios depend on pipe material and adjacent ground conditions. Examples of typical initial defect, deterioration, and collapse mechanisms are presented in Figures 2.2 and 2.3.

HYDRAULIC. Structural and hydraulic problems are closely related. Minor defects can lead to structural problems in specific soil conditions when a sewer is subjected to surcharge because of insufficient hydraulic capacity. A cycle of exfiltration and infiltration occurs that causes fines to migrate into the sewer, reducing lateral support from the soil. Eventually this leads to collapse of the conduit.

As sewers age, effective roughness increases and hydraulic capacity of the conduit drops, increasing the likelihood of surcharge and further structural deterioration.

Infiltration. Infiltration is water that enters a sewer system from the ground through defective pipes, pipe joints, damaged lateral connections, or manhole walls. Infiltration most often is related to a high groundwater level but can also be influenced by storm events or leaking water mains. Storm-influenced infiltration generally occurs in shallow sewers, such as house laterals in porous soils.

The rate of groundwater infiltration depends on the number and size of defects within a sewer and the head available. Typically, the rate of infiltration is greater during the winter or the end of the rainfall season and where pipelines are deepest. Groundwater flows may increase in highly permeable soils and significant quantities of groundwater can migrate through the granular bedding, which marks a relationship between soil type and infiltration rate.

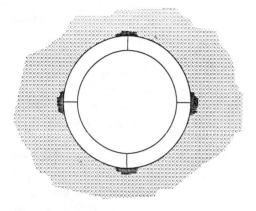

Stage 1:

Pipe cracking is caused by poor pipe laying practice or subsequent overloading or disturbance. The sewer remains supported and held in position by the surrounding soil.

Visible defects: cracks at crown, invert and springline. Infiltration may or may not be visible.

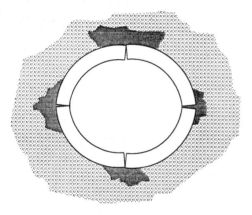

Stage 2:

Infiltration of groundwater or infiltration/exfiltration caused by surcharging of the sewer washes in soil particles. Side support is lost, allowing further deformation so that cracks develop into fractures.

Side support may also be insufficient to prevent deformation if the original backfill was either poorly compacted or of an unsuitable material.

Visible defects: fractures, slight deformation. Infiltration may or may not be visible.

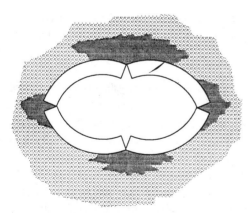

Stage 3:

Loss of side support allows side of pipe to move further outwards and the crown to drop. Once deformation exceeds 10%, the pipe becomes increasingly likely to collapse.

Development of zones of loose ground or voids caused by loss of ground into the sewer.

Visible defects: fractures and deformation, possibly broken.

Figure 2.2 Deformation of cracked pipes.

Groundwater levels can respond quickly to changes in tide-type fluctuations such as those experienced in coastal areas or along larger rivers, with

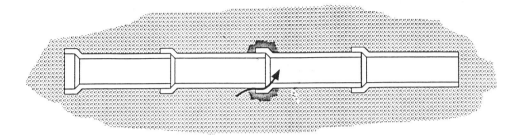

Stage 1: Gap in sewer at a joint or a poor lateral connection. *Visible defect*: Offset joint, badly made connection. Infiltration.

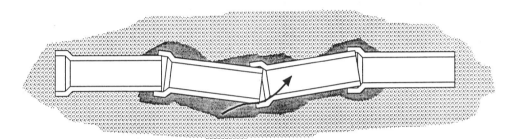

Stage 2: Infiltration of groundwater or infiltration/exfiltration caused by surcharging of the sewer washes in soil particles. Loss of soil support around the sewer allows pipe to move, opening joints and increasing the inwash of soil. *Visible defects*: open and displaced joints, loss of line and level. Infiltration.

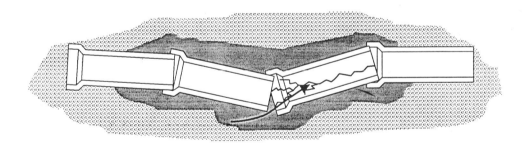

Stage 3: Uneven loading of pipes due to joint displacement causes cracking of pipes. Process then accelerates, and cracked pipes may also deform. *Visible defects*: Open and displaced joints, cracked and fractured pipes, loss of line and level. Development of zones of loose ground or voids caused by loss of ground into the sewer.

Figure 2.3 Subsidence of sewer.

the sewer being subjected to varying external head during the high-water cycle.

Inflow. Inflow is extraneous stormwater that enters a sanitary sewer system through roof leaders, cleanouts, foundation drains, sump pumps, and cellar, yard, and area drains. Stormwater may also enter through older connections between a sanitary sewer and storm sewers and through defective manhole covers and frame seals.

Connections usually can be located and the extent of the problem analyzed. Storm flow from impermeable areas draining to connections produce a certain flow during storms of known intensity, and annual flows of that connection to the total flow in a sewer can be determined.

In short, inflow reduces the hydraulic capacity of a sewer and increases the surcharging potential of the pipe, possibly contributing to sewer deterioration and increasing the potential for collapse. Infiltration contributes to sewer pipeline failures.

Typical system components that may be sources of infiltration and inflow (I/I) are shown in Figure 2.4. Attributing I/I to system components depends on an awareness of the physical characteristics of the system, including known defects, pipe and manhole condition, soil types, sewer depths, and length of sewers. All such conditions must be evaluated.

COMBINED SEWER OVERFLOW EFFECTS. Storm wastewater overflows are greatly influenced by the hydraulic performance of combined sewers and are a source of environmental pollution concern. Overflow settings

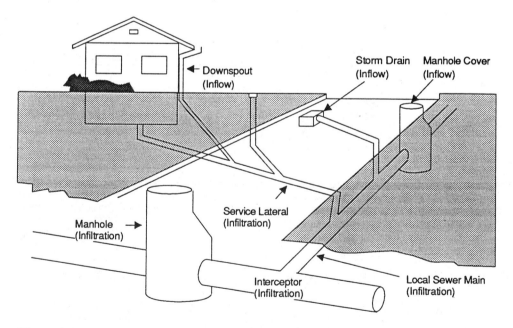

Figure 2.4   Collection system components and infiltration and inflow sources.

gradually become inappropriate because of increased runoff upstream caused by development in the drainage basin.

Combined sewer overflows (CSOs) may occur during and following storm events. Volume and frequency of CSOs are related to rainfall patterns in the given area served at discrete points in the collective system. Overflows contain constituents found in both urban stormwater and sanitary wastewater.

Flooding. Combined sanitary and stormwater sewers discharge wastewater to bodies of water during heavy rains. Sewers can also overflow onto land parcels or back up to building structures in the lower elevation portions of a drainage system. The type of discharge and level of flooding are related to the storm event and system operating characteristics.

Environmental. Combined sewer overflows often result in significant discharges to the receiving water of organic material, nutrients, sediment, microorganisms, oil and grease, metals, and other potentially toxic substances. Pollutant concentrations can be higher at the beginning of an overflow, called the "first flush" when material accumulated in a sewer is flushed out by the rapid increase in flow. Depending on the characteristics and sensitivity of the receiving water, an overflow can have serious effects.

CRITICAL SEWERS

Sewer systems have critical components whose failure would have particularly significant impact on the surrounding environment. After identifying those components, a sewer rehabilitation program is supposed to plan preventive maintenance and rehabilitation efforts toward them before addressing less critical sewers.

CATEGORIES. Sewers can be divided into three categories:

- Category A
    - Critical sewers where costs of failure would be high and effect on the surrounding environment great. Postfailure rehabilitation would be at least twice as expensive as planned replacement and three to four times the cost of rehabilitation.
- Category B
    - Less critical sewers because of reduced failure cost and effect, but where preventive action would still be cost effective.
- Category C
    - Noncritical sewers that have little or no effect as determined by critical definitions. Preemptive work in these sewers would not

be cost effective unless numerous failures occur in a confined area.

Critical sewers—those of Categories A and B—generally make up approximately 20% of a complete sewer system, of which 5% typically would be Category A critical sewers. Wide variations may occur from this distribution, depending on the characteristics of individual systems (*Sewerage Rehabilitation Manual*, 1986).

A rehabilitation program should keep Category A sewers failure free and reduce the failure rate of Category B sewers, and full inventory of pipeline characteristics is recommended for both categories. Pipe characteristics include general data on the location, diameter, slope, depth, construction year, material, condition, use, traffic loading (vehicles per day), soil data, water table level, I/I quantity, and major repair or renovation history.

CRITICAL DEFINITIONS. To determine the category and critical nature of a sewer, critical sewer definitions pertinent to the particular system must be developed. Identification of critical sewers is usually a one-time task and forms the foundation of a preventive approach to future rehabilitation.

Critical sewers are likely to meet one or more of the following conditions:

- Old age (older than 50 years);
- Deeper than average;
- Large diameter;
- Brick or masonry construction;
- Unlined or unreinforced concrete;
- Located under busy streets, in streets providing access to emergency service, or in sensitive environmental locations (offshore, wetlands);
- Difficult access (such as in easements or under structures);
- At or near hydraulic capacity and in poor soil conditions or high groundwater;
- Difficult to bypass;
- Require frequent inspection or maintenance; and
- Excessive I/I.

Criteria specific to a system must be established before screening can take place, and an objective, disciplined procedure for analyzing the system must be adopted. The critical sewer identification process is presented in Figure 2.5.

After determining what proportion of a sewer system is critical, critical pipe segments can be identified using the methodology described above. Preventive maintenance and rehabilitative efforts should be focused on the most critical sewers first to ensure the most cost-effective use of resources.

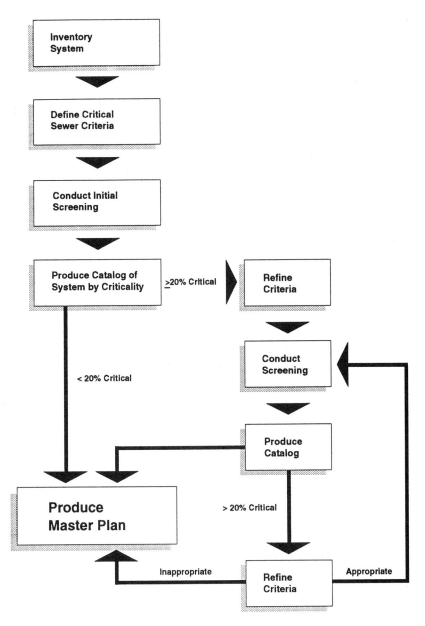

Figure 2.5 Critical sewer identification process.

*P*ERFORMANCE CRITERIA

Engineering and procedural steps are required by regulatory agencies in design and construction of sewerage projects. These criteria apply to all facilities proposing wastewater treatment and discharge. The primary reason for ensuring that plans conform to engineering standards is to protect public health.

Structural and hydraulic performance of sewers influence the integrity and service life of conduits. The cost of sewer failure would be significantly higher than the cost of rehabilitation, and hydraulic upgrading generally is aimed at areas where flooding and pollution would be severe in the event of sewer failure.

Standards for inspection and engineering design and those for performance standards of sewerage works are uniform across the U.S. Design standards are not consistent throughout the world, however. Individual governing agencies are responsible for setting standards of collection, treatment, and disposal system performance.

With input from regulatory agencies, performance criteria might include the following:

- Public health criteria that would differ, depending on what land area would be affected by flood waters. Specific criteria might depend on the type of effluent, degree of flooding, and whether a structure was residential, commercial, or industrial.
- Structural integrity so a sewer could be measured by a design factor of safety against failure. Integrity could be measured by assessing the risk of failure where a sewer has deteriorated or corroded from existing structural defects.
- Receiving water quality such that the major impact of sewerage systems on water quality is linked to CSOs. Operation of CSOs should not compromise receiving water quality with respect to bacteriological and aesthetic impacts.

Different system performance levels can be established, depending on regulatory requirements, regional values, and costs. Criteria can be either fixed or variable. Fixed performance criteria are appropriate under one or more of the following conditions:

- There is no readily available basis to assess performance;
- Performance levels are evaluated free of pressures such as political concerns;
- Small increments in performance cause small benefit costs; and
- A simple approach is needed.

Variable performance criteria are appropriate under one or more of the following conditions:

- Real performance can be anticipated accurately;
- Other conditions add to the costs, such as social considerations;
- Different levels of performance have different cost implications; and
- Achieving a preferred performance level in one location would be too costly in another.

*I*NTEGRATED APPROACH

The rehabilitation strategy for any facility, ranging from individual pipeline needs to the complete collection system program approach, involves investigation of both structural condition and hydraulic performance. This leads to a long-term planning approach that encompasses the overall needs of the system.

Corrosion is often the cause of failed or failing pipeline, manholes, or structures, and failures for this reason should be evaluated historically. Corrosion normally is most destructive during a system's early use before it is subjected to the velocities and flows that normally inhibit corrosion. An evaluation of pipeline or facility condition should identify structural and corrosion damage and assess its severity and the potential consequences of failure.

NEED. Understanding hydraulic and structural performance of a sewer system is paramount to its cost-effective management. This approach involves investigative effort illustrated in Figure 2.6 as a flow chart divided into four phases, each of which contains steps to conduct the work (*Sewerage Rehabilitation Manual*, 1986, and Schrock, 1991).

PLANNING INVESTIGATIONS (PHASE 1). This phase deals with work before a detailed investigation of a sewer system.

Review System Records. Determine information for identifying critical sewers from existing system maps and records.

Categorize Sewers. Rate sewers relative to their importance to system operation and their potential effect in the event of failure.

Document Performance. All records of stoppages, collapses, spills, crisis responses, and damage usually available from the U.S. Environmental Protection Agency (U.S. EPA) should be studied. The extent and frequency of problems and spills and collapses should be reviewed in detail to establish if there is evidence of potential problems. This would provide information for initiating a rehabilitation or replacement program.

Select a Detailed Investigation and Establish Priorities. A detailed investigation is justified when critical sewers are or are suspected to be hydraulically overloaded or where corrosion or structural problems are possible and rehabilitation is restrained by insufficient capacity. Where such conditions are known not to exist, a cursory investigation may be sufficient.

Improve System Records and Access. A detailed investigation requires the use of resources that typically are limited. To make the best use of available resources, it will be necessary to provide sufficient information for hydraulic,

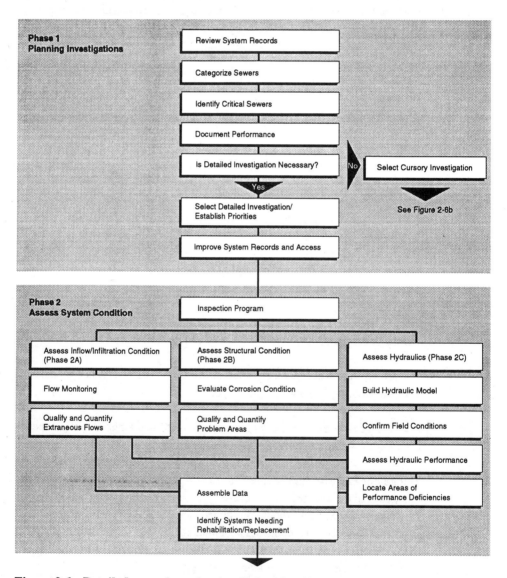

Figure 2.6 Detailed sewer investigation (Schrock, 1991).

corrosion, and structural investigations. Any improvements to records or maps should be coordinated with needs of the investigation.

ASSESSING THE SYSTEM CONDITION (PHASE 2). Assessment of the system condition involves an inspection to determine the I/I condition, structural condition, and hydraulics. This step should include the selection of

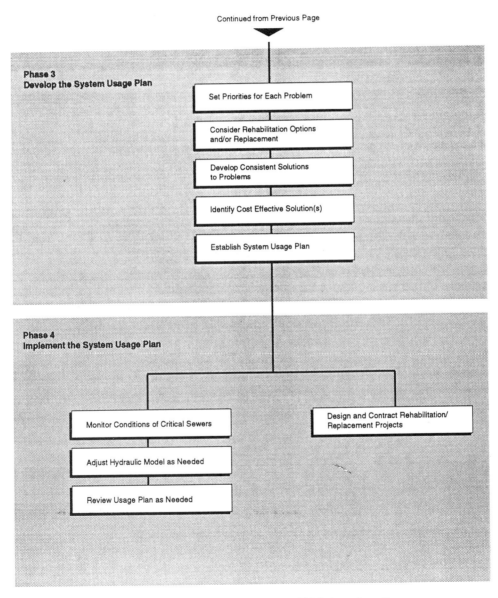

Figure 2.6   Detailed sewer investigation (Schrock, 1991) (continued).

General Approach

certain sewers for closed-circuit television (CCTV) viewing or entry inspection and survey work, as discussed under "Assessing Hydraulics."

ASSESSING THE INFILTRATION AND INFLOW CONDITION (PHASE 2-A).
Phase 2-A assessment involves evaluating extraneous flows in a sewer system, including infiltration from groundwater levels or rainfall and inflow from storm drainage sources that enter the system through cracks, displaced joints, leaking, porous walls, voids, and broken or illegal service connections.

Infiltration and Inflow Condition. Existing CCTV video or entry information on the selected sewer should be reviewed to provide information beneficial in determining points of extraneous flow and identifying mainline and service connection problems.

Flow Monitoring. Measuring devices in the system should be correlated to determine dry and wet weather flows. Measuring records are important in isolating and reducing problems to as small an area as possible. A temporary flow-monitoring program should be initiated through installing measuring devices at selected locations (see Chapter 6).

Qualify and Quantify Extraneous Flows. The measurement data will require analysis to develop normal system flow. Total flow after a rainfall event will include groundwater infiltration (GWI), stormwater inflow (SWI), and rainfall-dependent infiltration (RDI), and such sources must be quantified to determine normal sanitary wastewater system flow. Normal (or dry weather) flow typically is determined first and then subtracted from wet weather flow to determine GWI, SWI, and RDI quantities. Applicable flow-monitoring data should be used when assessing hydraulic performance.

ASSESSING THE STRUCTURAL CONDITION (PHASE 2-B).
Phase 2-B assessment involves the inspection of critical sewers and the identification of rehabilitation, replacement, or future condition-monitoring requirements.

Structural Condition. The CCTV video or entry information from the selected sewers should be reviewed for structural problems such as cracks, breaks, displaced joints, missing pipe pieces, clay liners, roots, sags, and corrosion.

Evaluate Corrosion. Corrosion of a cementitious material will continue unless the corrosive environment is eliminated or the corrodible material is suitably protected. Thus, corrosion must be identified and an assessment made of its progression so remedial action can be taken before failure occurs.

Qualify and Quantify Problem Areas. A rating system should be established to identify the failure mode—structural or corrosion—and its magnitude, condition, and progress. This is discussed in Chapter 3.

ASSESSING HYDRAULICS (PHASE 2-C). Phase 2-C assessment involves the investigation of the hydraulic performance of the sewer system in general and the critical sewers in detail to establish their performance and any needs for improvement.

Hydraulics. The actual flow rate under varying field conditions must be determined to correctly evaluate the system.

Build Hydraulic Model. A hydraulic model of the system on which performance and needed improvement can be based should be developed, especially for longer systems. Critical sewers and the collection system should be modeled in detail, and this data incorporated into a suitable computer model. Modeling of smaller, less complex systems also is often beneficial.

Confirm Field Conditions. To develop input as the basis for obtaining correct hydraulic information, actual field data is needed from maps and construction drawings. Actual invert elevation and pipe diameters typically must be verified in the field.

Calibrate Model. The model is calibrated against actual flow data from a properly conducted monitoring program (Chapter 6) to confirm model accuracy and validity. Adjust model and recalibrate as required.

Assess Hydraulic Performance. The model is run with a range of flow conditions to establish when each sewer and ancillary structure reaches appropriate anticipated design criteria.

Locate Areas of Performance Deficiencies. Sewer reaches that fail to meet minimum required hydraulic performance criteria are identified, and the principal causes established.

Assembling Data. Information obtained in the I/I, structural, and hydraulics assessments is assembled and evaluated.

Identifying Systems Needing Rehabilitation or Replacement. This step is to identify components of the system in need of work and what is required to bring the system up to its level of required performance, including point repair, rehabilitation, or replacement.

DEVELOPING THE SYSTEM USAGE PLAN (PHASE 3). Phase 3 involves evaluating all the identified conditions needing correction (excess I/I, structural duress, and hydraulic deficiencies), assessing potential solutions, and producing a plan to accomplish improvements in the most cost-effective way.

Setting Priorities for Each Problem. The system usage plan should incorporate all existing and predicted needs, though limited funding may keep some from having immediate priority. Projects should be completed in stages to facilitate planning within funding constraints.

Considering Rehabilitation Options or Replacement. All possible solutions to problems must be considered, taking into account scheduling and coordinating the work.

Developing Consistent Solutions to Problems. To expedite design and construction, a format for developing consistent solutions to problems must be used.

Identifying Cost-Effective Solutions. To identify the most cost-effective combination of solutions on which the system usage plan can be based, an assessment is needed of rehabilitation or replacement costs, expected longevity and performance, and any additional factors such as social implications, safety, maintenance and operation costs, and reliability.

Establishing the System Usage Plan. When preferred solutions are identified, final planning involves scheduling various proposed projects and the requirements for future monitoring of I/I and structural condition of the critical sewers. The plan should also include project activities such as relief lines, interconnections, and bypassing that affect the critical sewer components.

IMPLEMENTING THE SYSTEM USAGE PLAN (PHASE 4). Phase 4 consists of implementing the system usage plan and monitoring the system to ensure the plan is appropriate or modified as necessary.

Designing and Constructing Rehabilitation and Replacement Projects. This involves developing individual projects in the system usage plan through detailed design and construction, applying the established design criteria. The timing should be coordinated with other project work in the system, updated as necessary by subsequent findings, and scheduled to be accomplished with available funds.

Monitoring Conditions of Primary Sewers. Infiltration and inflow reduction and the structural condition of critical sewers should be monitored regularly to confirm that assessed needs and priorities are being satisfied.

Adjusting the Hydraulic Model as Needed. So it can be used reliably for design work, to assess planned and unplanned changes, and to verify actual results, the hydraulic model must be kept up-to-date so it continues to represent ongoing adjustments in the system.

Reviewing Usage Plan as Needed. Sewer maps and records should be updated and the usage plan reviewed on an ongoing basis as events occur or circumstances change that affect future projects. Such changes should be noted and modifications made to the usage plan.

OUTLINE OF CURSORY INVESTIGATION. The cursory investigation involves the same four distinct phases and has similar steps as the detailed investigation, although at a reduced level. These steps are shown in Figure 2.7.

PREDICTIVE MODELING

Mathematical modeling is a tool used to define the conditions under which a sewer system must operate and its performance under those conditions. Properly constructed and calibrated mathematical models can predict runoff rate and volume entering storm or combined systems, infiltration entering any system, and the consequences of, for example, surcharge and overflow or street flooding. Available models cover a wide range of complexity and function. The choice of model for any particular problem depends on the detail required and cost involved (more complex models cost more to develop and use).

NECESSARY FUNCTIONS. Mathematical models need to be able to convert rainfall information to sewer flow (hydrologic function) and, subsequently, route that flow through the defined sewer network (hydraulic function).

Hydrologic Function. Almost all models of this type are capable of hydrologic calculations to convert rainfall on pervious and impervious surfaces to surface runoff. Necessary algorithms are well known, and several acceptable formulations are used.

A few models are capable of simulating snow melt in addition to runoff from rainfall. In many areas of the U.S., the combination of snow melt and

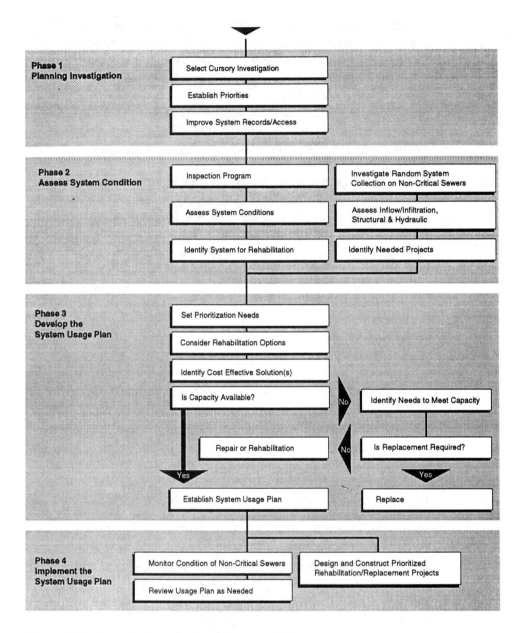

Figure 2.7 Cursory investigation (Schrock, 1991).

rainfall may induce maximum I/I. Introduction of this complexity requires that air temperature be included in the simulation.

Unlike surface runoff, few models can compute infiltration from applied rainfall. A few (for example, version 4.0 of the EPA Storm Water Management Model [SWMM]) are capable of simulating the flow of groundwater to sewers by performing a mass balance on groundwater flow. Others provide functional relationships between rainfall over a given period and infiltration. These approaches are limited in their ability to simulate short-term infiltration processes.

Infiltration is a result of increases in the groundwater table that vary over the short term (days) and over the annual cycle. There is also a short-term effect within hours of rainfall because of subsurface interflow or the behavior of sewer trenches as french drains.

Successful correlation of rainfall to infiltration has been achieved by relating infiltration to antecedent precipitation indexes with time scales of hours to several months and by linearly adding effects in a unit-hydrograph-like approach. The antecedent precipitation index is a measure of soil moisture conditions defined by $P_a = P_1 + P_2/2 + P_3/3+....+ P_t/t$ where P_t is the precipitation that occurred t time periods in the past.

Hydraulic Function. Hydraulic calculations in available models range from simple accounting of time of travel to solution of full equations of motion. Simplified methods, such as the kinematic wave approximation, route hydrographs by time translation but retain the hydrograph shape. These simplified approaches can neither account for backups induced by downstream restrictions nor provide data on the degree of surcharge or increase inflow available with surcharge.

Models that incorporate full dynamic equations of motion (such as the SWMM EXTRAN Block) are needed when degree of surcharge information is required, where the increase of flow capacity with surcharge is important, and where complicated structures whose behavior depends on flow depth must be simulated. Increased complexity necessitates a corresponding increase in modeling effort and cost.

MODELING TIME SCALE. Most available models are designed principally for examination of a limited time period, usually a single rainfall event. This time scale is appropriate for examination of large design events to determine system capacity requirements.

To evaluate structural condition of sewers, it is necessary to consider events that occur once to several times per year. Sewers that may be surcharged with high frequency are at special risk because the surcharge may pump the surrounding soil into the sewer, creating voids and reducing the exterior support structure. Sewer flows from a long-term rainfall record (10 to 15 years or more) must be simulated to determine the frequency of surcharge. By accounting for antecedent precipitation, this yields a true flow frequency (rather than the rainfall frequency), thus an appropriate surcharge frequency. Where snow melt is an important determinant of flow extremes, temperature is added to the simulation. Diurnal and weekly variations in wastewater flow are also considered.

JUSTIFICATION FOR MODELING. The proper use of mathematical modeling techniques can provide the user with an accurate picture of sewer system performance. For example, flow-monitoring studies usually cover a limited period during which comprehensive conditions may or may not be observed. Using mathematical modeling calibrated and verified by a carefully planned and executed flow-monitoring program to extend the performance evaluation to a long-term rainfall record provides data on expected frequencies, such as how often a sewer is overloaded, which is more accurate than the use of synthetic design rainfall simulation. Properly calibrated modeling also provides a complete system picture in which upstream and downstream effects are integrated (for example, improving capacity upstream may overload downstream sewers).

Modeling is also useful in examining effects before and after rehabilitation. Determination of the effect of rehabilitation for I/I control is a prime example. It is difficult but possible to accurately define I/I reduction from flow monitoring before and after rehabilitation work. Antecedent rainfall and groundwater conditions may be different and must be considered in the determination. With modeling, however, the model can be calibrated with "before" and "after" flow-monitoring data, and both models can be used to simulate a long-term rainfall record to generate flow statistics.

Improperly calibrated models may result in either under- or overestimation of flows and rehabilitation effectiveness by several orders of magnitude with corresponding overspending or underfunding to meet actual requirements.

REFERENCES

Institution of Civil Engineers (1981) Restoration of Sewerage Systems. In *The Sewer Dereliction Problem*. Proc. Int. Conf. Inst. Civ. Eng., London, U.K.

Schrock, B.J. (1991) Pipeline Systems Rehabilitation Workshop. San Jose, Calif.

Sewerage Rehabilitation Manual (1986) (Addenda published in 1990). 2nd Ed., Water Res. Cent., Eng., U.K.

Urban Institute (1981) *Capital Infrastructure Investment Needs and Financing Options*. HUD-0003404, Dep. Housing and Urban Dev., Washington, D.C.

Chapter 3
Methods of Structural Evaluation

27 **Pipeline Conditions**
28   Structural Conditions
28   Corrosion Conditions
28   Qualification and Quantification
28   Brick Sewers
29   Concrete and Clay Sewers
30   Rehabilitation Assessment

32 **Program Plan**
32   Evaluation Assessment
33   Monitoring
34   Stabilization
34   Rehabilitation
35   Replacement
35 **References**

Structural condition assessment is a principle objective of any pipeline system inspection program. Closed-circuit television (CCTV) or entry information of the selected sewer segment requires careful review and analysis to identify where structural rehabilitation or replacement is required.

Field inspection provides information on the corrosion or deterioration of a cementitious or corrodible material. This also provides information about specific location conditions that affect the hydraulic performance of individual pipeline reaches, such as sediment, debris, roots, open joints, and misaligned joints.

PIPELINE CONDITIONS

In addition to revealing opportunities for correcting capacity needs, pipeline evaluation should identify structural and corrosion defects requiring correction, their severity, and the potential consequence of failure. Pipeline condition assessment helps establish priorities for rehabilitation or replacement. The likelihood of failure and the associated risk analysis are intrinsic to the evaluation when budgetary constraints affect the work (*Sewerage Rehabilitation Manual*, 1986).

STRUCTURAL CONDITIONS. Sewer pipeline can become impaired for a number of reasons. Pipeline repair can be accomplished more readily when the cause of failure is determined.

Various types of pipes are prone to certain types of failure based on their physical design. In older pipe designs, for example, cement-packed or grouted joints tend to deteriorate over time. Some joints are out-of-round, permitting root penetration followed by cracking and infiltration; some were cracked or broken when installed, permitting similar problems; some were designed or installed without proper joint articulation near structures or other differential settlement areas; some were installed without proper bedding before backfilling and developed pipeline sags; and some pipeline joints were improperly grouted, packed, tarred, or gasketed.

Generally, any type of pipe when improperly designed or installed may cause structural failure or be prone to infiltration. Improper connection of service laterals may cause structural problems to the main line. Locating the deficiency and determining its cause permit proper correction. Older vitrified clay pipe may have more porous walls because it was not highly vitrified.

CORROSION CONDITIONS. When pipeline, manholes, or structures have failed or are failing from corrosion, their history should be inspected. As velocities and flows normally inhibit corrosion, corrosion is usually most active during a system's early use, when low flows occur before design discharges.

Two important aspects of corrosion evaluation are physical or CCTV inspection of the sewer lines and the calculation of a sewer's remaining life expectancy. This work is covered in detail in the American Society of Civil Engineers manual *Sulfide in Wastewater Collection and Treatment Systems* (1989).

QUALIFICATION AND QUANTIFICATION. A rating system must be established for identifying the structural or corrosion failure mode and the failure state, magnitude, conditions, and progress. To evaluate and compare the pipeline condition, various assumptions must be made regarding characteristics that indicate potential failure (see Chapter 2).

BRICK SEWERS. The first deterioration that occurs in brick sewers is general mortar loss by erosion, corrosion, or aging. In many sewers, the cement between bricks has entirely degraded to a layer of loosely bound sand particles, and the mortar can easily be removed from between the bricks. Nevertheless, many such sewers have maintained their original cross-sectional shape and, as long as the deteriorated mortar remains in place, the sewer structure will support the soil loadings. The structural danger is that the softened mortar may be eroded by high flows in the sewer or by groundwater infiltration. The soft mortar also offers little resistance to root intrusion, which can accelerate mortar loss, produce structural deformation, and obstruct flow.

When a sewer loses a significant portion of the mortar between bricks, it begins to deform under the soil loads. Complete loss of the mortar results in loosened and, eventually, missing bricks, allowing severe deflections. The sewer may also deform vertically by developing longitudinal cracks on the crown either before or after the mortar is severely degraded. Large deformations or extensive cracking are the final prefailure warnings in brick sewers.

Brick sewer structural condition evaluation criteria are presented in Table 3.1. The relative importance of the criteria is indicated by their position on the list, with the most critical listed first, indicating the greatest structural deterioration.

CONCRETE AND CLAY SEWERS. Evaluation criteria for the structural condition of concrete and clay sewers are less complex than those for brick

Table 3.1 Brick sewer structural condition evaluation criteria.*

Structural condition	Evaluation criteria
Sags	The pipeline invert drops below the downstream invert.
Vertical deflection and cracks	Vertical dimension of the sewer is reduced. Crack lines are visible in the brickwork or mortar; bricks have moved apart from one another. Bricks are still in place.
Missing bricks	Single bricks or areas of bricks are missing. More than one ring may be affected.
Lateral deflections	Sewer is deformed or original cross section of sewer is altered.
Root intrusion	Tree or plant roots have grown into or entered the sewer through the brick interstices.
Missing mortar	Mortar between brickwork is missing to a degree varying from surface loss to medium or total loss. Bricks are still in place.
Loose bricks	A forerunner of missing bricks is displaced bricks, that is, single bricks or areas of bricks have moved from their original position.
Protruding lateral	A service outlet or pipe section protrudes or extends into the sewer, varying in magnitude.
Soft mortar	A forerunner of loose or missing brick and sewer shape change is soft mortar, usually caused by corrosion.
Depth of cover	The deeper and larger the sewer, the more critical the potential problem.

* Other criteria may be incorporated.

sewers. Nevertheless, there are more criteria to evaluate primarily because of the greater diameter and size range of the sewers involved.

Many concrete and clay sewers continue to function even with a critical structural condition. However, the pipeline will continue to deteriorate, depending on the condition and the internal or external environment. The soil structure above the pipe will eventually collapse, depending on the magnitude of the problem. These criteria are listed in Table 3.2.

Proper use of structural condition evaluation criteria is the system designer's responsibility. A numerical importance rating factor should assigned to each criterion. Internal condition rating factors are presented in Table 3.3.

Depending on the extent of the condition throughout a given pipeline reach, a minor, moderate, or severe multiplier factor is used, such as 1, 2, or 3, respectively. Also, the effects of external factors such as soil types, surcharge, water table and fluctuation, and traffic are used.

*R*EHABILITATION ASSESSMENT

The choice of a proper, cost-effective rehabilitation procedure for a given sewer is best made through a thorough understanding of all possible methods along with knowledge of pipeline conditions. A complete evaluation is essential in determining whether it is more cost effective to rehabilitate or replace a particular pipeline reach or if merely monitoring the pipeline is the correct choice.

The wide range of old piping materials and their respective joints makes it impossible to generalize about how effective particular rehabilitation methods are or whether rehabilitation is practical without considering the age and durability of the materials. Any decision to replace or rehabilitate a particular pipeline will depend on the performance to be gained from rehabilitation compared to replacement with modern materials using correct installation practices.

Rehabilitation methods, other than joint grouting or sealing, reduce pipe section and can affect hydraulic performance. While often acceptable, such reduction must be considered in the rehabilitation evaluation. Where major pipeline structural strengthening is required, the sacrifice of hydraulic for structural performance can be significant.

An understanding of the mechanisms of structural failure is essential to selecting effective rehabilitation techniques. There are several factors involved in failure, the more significant being

- Corrosive soils or groundwater;
- Voiding of bedding and backfill as groundwater enters or weakens a failing pipe;
- Sulfide formation in wastewater;

Table 3.2   Concrete and clay sewer structural condition evaluation criteria.*

Structural condition	Evaluation criteria
Collapsed pipe	There is complete loss of structural integrity of the pipe. Most of the cross-sectional area is lost.
Structural cracking with deflection	Pipe wall is displaced.
Longitudinal	Defect runs approximately along the axis of the sewer.
Circumferential	Defect runs approximately at right angles to the axis of the sewer.
Multiple	Longitudinal and circumferential defects are combined.
Slab-out	A large hole in the sewer wall exists with pieces missing.
Sag	The pipeline invert drops below the downstream invert.
Structural cracking without deflection	Sewer wall is cracked but not displaced, longitudinally, circumferentially, or multiply.
Cracked joints	The spigot or bell of a pipe is cracked or broken.
Open joints	Adjacent pipes are longitudinally displaced at the joint.
Holes	A piece of pipe wall or joint is missing.
Root intrusion	Tree or plant roots have grown into or entered the sewer through an opening in the pipe wall.
Protruding joint material	The original joint sealing material is displaced into the sewer from its original location.
Corrosion	The cementitious pipe material shows evidence of deterioration, illustrated by the following stages.
Stage 1	The pipe wall surface shows irregular smoothness, that is, wall aggregate is exposed.
Stage 2	The reinforcing steel is exposed.
Stage 3	The reinforcing steel is gone or the pipe wall is no longer intact, revealing the surrounding soil.
Pulled joint	Adjacent pipe joints are deflected beyond allowable tolerances.
Protruding lateral	A service outlet or pipe section protrudes or extends into the sewer, varying in magnitude.
Vertical displacement	The spigot of the pipe has dropped below the normal joint closure.
Depth of cover	The deeper and larger the sewer, the more critical the potential problem.

* Other criteria may be incorporated.

- Corrosive and erosive industrial effluent;
- Root intrusion at deteriorated joints or pipe barrel;
- Differential settlement of soil adjacent to the pipe;

Table 3.3 Internal condition rating factors.

Description	Rating factor
Collapse or collapse imminent	5
Collapse likely in forseeable future	4
Collapse unlikely in near future, deterioration likely	3
Minimal collapse risk in short term but potential for further deterioration	2

- Loading in excess of design limits; and
- Improper installation during original construction.

A pipeline investigation should help determine possible causes of failure, and eliminating causes will reduce the list to a manageable size for continued study. Soil borings, wastewater analysis, pipe crown or manhole wall pH determination, and pipe wall structural and pipe or soil interaction analysis provide technical information needed to develop a conclusion.

Program Plan

When evaluation and physical assessment have been completed, rehabilitation begins. Differences in the rates at which sewer system components deteriorate are used to establish priorities of rehabilitation actions.

Some rehabilitation techniques are more effective in dealing with certain types of failure than others, thus rehabilitation that addresses failure symptoms without considering failure mechanisms invites costly errors. Through a staged screening process, the applicability of repair methods should be assessed in relation to special construction problems, sizes and shapes of existing pipes, pipe conditions to be corrected, and desired capacities of the rehabilitated system. Significant differences in the capabilities of rehabilitation methods will significantly reduce the field of alternatives.

Evaluation Assessment

The goal of sewer rehabilitation is to arrest deterioration. Rehabilitation ranges from repairing or stabilizing existing pipe or pipeline to major in-place construction of linings or full replacement.

There is the option of deferring rehabilitation until the risk of collapse is unacceptable, though this point varies because acceptable risk depends on factors external and internal to the sewer. Though option is viable because sewers do not normally fail without first showing signs of distress, it requires scheduled monitoring of the progress of deterioration. Expenditure of construction funds is delayed until there is a risk of losing rehabilitation options through increasing pipeline degradation.

The choice of rehabilitation approach should be based on the information available on the sewer system and on the financing available for the project. Based on the information initially developed, the following alternatives are proposed for evaluation:

- Level 1—monitoring and information collection;
- Level 2—stabilization of existing sewer;
- Level 3—rehabilitation of existing sewer; and
- Level 4—replacement of existing sewer.

MONITORING. This alternative involves additional information to support a decision to either rehabilitate or stabilize an existing sewer. By verifying the quality of the piping and the competence of the soil around the sewer, a program can be formulated that at least stabilizes a sewer. Conversely, if additional data collection reveals more serious problems, a higher level of rehabilitation may be needed.

A routine inspection program should be established to monitor performance and develop needed design information that could help cut rehabilitation costs.

For critical sewers, follow-up inspections should be made approximately 1 year after the initial inspection, unless a level 3 or 4 program is initiated. Inspection also is required to verify the construction integrity of a lined or replaced sewer, though intervals between inspections can increase as confidence in the sewer's condition increases. Subsequent inspections can be made every 2 to 3 years, depending on initial and subsequent findings. Frequent initial inspections provide an early indication of change and opportunities to adjust the monitoring program or the decision-making process.

This approach recognizes that the pipe will reveal conditions that may indicate further degradation or excessive distress in the sewer. Although regular sewer inspections should be conducted regardless of the construction material, those systems known to have problems or with unlined concrete pipe, cracked clay pipe, and loose brick are more critical.

Monitoring and research should provide the following pipeline information:

- Failure mode
  - Corrosion
  - Structural
  - Pipe deficiency
  - Joint leaks

- Failure state
  - Magnitude
  - Condition
  - Progress
- Soil condition
  - Soil type (pipe zone)
  - Soil type (backfill)
  - Groundwater

The age of the sewer and various operating procedures must also be considered in the monitoring program.

STABILIZATION. Point repair or stabilization of the existing sewer can extend the useful life of the pipeline. Point, or spot, repair is the replacement of a collapsed or seriously fractured pipe length between manholes instead of total pipeline rehabilitation or replacement. When a point repair program is implemented, additional repair is needed to correct new defects discovered during subsequent inspections. It is normally more cost effective to proceed with rehabilitation when conducting the initial point repair for the entire pipeline if it is warranted.

An alternative involves modifying and stabilizing soil around the pipeline, which is intended to restore soil competence around the sewer and produce uniform circumferential loading that favors the strengths of a pipe. Cement or chemical grout stabilization could be appropriate with either stabilization or rehabilitation.

REHABILITATION. When suffering from corrosion, breaks or fractures, unsound materials, or other signs of excessive loading or deterioration, pipelines require repair measures beyond mere stabilization. Rehabilitation is the next level of effort and involves various lining systems constructed within the existing sewer.

Rehabilitation can provide corrosion protection and certain levels of tolerance under structural duress. A design using some rehabilitation methods can withstand external and internal loading conditions considering the remaining structural value of the existing reinforced concrete or vitrified clay pipe. Pipeline size and environmental factors such as accessibility and effect on current service requirements influence the choice of a rehabilitation method. Pipe insertion or sliplining is well suited for common pipeline sizes and shapes, though not all types of pipeline problems can be rehabilitated by this method. Pipeline capacity, size, shape, and condition may permit or require other choices, including custom materials and rehabilitation methods.

Criteria used to select acceptable rehabilitation methods based on the condition of the existing pipe and pipeline can be divided into three categories.

Condition I applies to unprotected concrete surfaces with surface corrosion and exposed aggregate or minor cracks or fractures having no pipe wall displacement. If joints are leaking or open with minor displacement, chemical grouting or mechanical sealing devices may be used. Point repair normally is used at needed locations. Total replacement usually is unnecessary at this point.

Condition II applies to unprotected concrete surfaces with corrosion where steel reinforcement is exposed. Also, cracks or fractures having minor displacement may be rehabilitated. Point repair normally is used at needed locations, and total replacement may be desirable.

Condition III applies to unprotected concrete surfaces with corrosion where steel reinforcement is missing or holes through the wall exist. Also, cracks or fractures having serious displacement and holes or slab-outs occurring in the wall typically can be rehabilitated. The existing pipeline will require pulling a television camera or slipliner proofing section though the line to verify available cross sections to rehabilitate. Point repair often is required in addition to total rehabilitation, though total replacement may be preferable. Life cycle and cost effectiveness should be evaluated at this time.

REPLACEMENT. When pipelines are found to be beyond repair using any of the rehabilitation methods, total replacement is recommended. Normally, this condition exists at specific locations, and point repair or rehabilitation may be the most cost-effective measure for the remaining system. Where pipelines are determined to be hydraulically overloaded and upgrade rehabilitation will not improve the flow condition, pipeline replacement or relief line construction is recommended.

REFERENCES

American Society of Civil Engineers (1989) *Sulfide in Wastewater Collection and Treatment Systems*. Manual of Practice No. 69, New York, N.Y.

Sewerage Rehabilitation Manual (1986) (Addenda published in 1990). 2nd Ed., Water Res. Cent., Eng., U.K.

Chapter 4
Methods of Infiltration and Inflow Evaluation

38	Purpose	53	Data Recording
40	Problems	53	Building Inspection
40	Evaluation Techniques	54	Building Inspection Program Procedure
40	Flow Components		
41	Peak Rate Determinations	55	Dye-Water Testing
41	Annual Volume Determination	56	Determining Conditions Caused by Storm Sewer Sections
42	Preinstallation Considerations		
43	Meter Maintenance	56	Determining Conditions Caused by Stream Sections
44	Data Evaluation		
44	Precipitation Measurement	56	Identifying Sources on Private Property
45	Groundwater Gauging		
46	Smoke Testing	56	Identifying Structurally Damaged Manholes
47	Procedure		
49	Photographs	57	Verifying Sources Found by Other Means
49	Manhole and Pipeline Visual Inspection		
		57	Safety Measures
49	Inspection	58	Logging Test Results
50	Safety Measures	58	Night Flow Isolation
51	Atmospheric Hazards	59	Groundwater Migration
51	Explosive Gases	60	Single-Pass Method
51	Toxic Conditions	60	Multipass Method
51	Oxygen Deficiency	60	Isolation Techniques
51	Physical Injury	60	Plugging
52	Infections	62	Differential Isolation
52	Animals	62	Portable Weir Methods
52	Chemicals	63	Velocity-Area Method
52	Drowning	64	Fluorometric Methods

64	Results	72	Cost Effectiveness
65	Pipeline Cleaning and Television Inspection	73	Evaluation of Inflow
		74	Applicable Techniques
66	Photographs	74	Methodology
66	Leak Quantification	77	Rainfall-Induced Inflow Localization
68	Evaluation of Infiltration	78	Cost Effectiveness
68	Water Use Evaluation	79	Evaluation of Physical Conditions
69	Maximum–Minimum Daily Flow	79	Structural Integrity
69	Maximum Daily Flow Comparison	80	Operational and Maintenance Problems
70	Nighttime Domestic Flow Evaluation		
70	Small Subsystems	80	Map Correction and Updating
71	Infiltration Location	81	Applicable Techniques
72	Groundwater Levels	81	Visual Inspection
72	Interviews	82	Television Inspection
72	Visual Inspection	83	References

Serious problems can result from excessive infiltration to sewers from groundwater sources and from high inflow rates to sewer systems directly from sources other than those that sewer conduits are intended to serve. The effects of these extraneous flows are particularly important because urban growth creates the need for all available sewer system capacity. The effects of bypassed, spilled, and undertreated wastewater caused by infiltration and inflow (I/I) are deterrents to the overall objective of protecting the nation's water resources.

*P*URPOSE

Since the early 1970s, a determined effort has been made in the U.S. to reduce the effects of I/I on sewer systems. To assist engineers, municipalities, and regulatory agencies, the U.S. Environmental Protection Agency (U.S. EPA) prepared a two-part set of guidelines for conducting I/I evaluations: the I/I analysis and the sewer system evaluation survey (SSES).

The I/I analysis is intended to detect any excessive infiltration and inflow. The SSES provides a detailed evaluation and corrective action plan for areas where the I/I analysis determines that excessive I/I exists. Correction of infiltration to existing sewer systems involves: (1) evaluation and interpretation of wastewater flow conditions to determine the presence and extent of excessive extraneous water flows from sewer system sources; (2) the location and gauging of such infiltration flows; (3) the elimination of these flows by various repair and replacement methods when cost effective; and (4) a diligent, continuous maintenance program.

In the case of inflow, the problem also includes prevention and cure. Prevention of excessive inflow is a matter of regulating sewer uses and enforcement of applicable precepts and codes by means of information obtained through surveys and surveillance. Correction of inflow problems involves discovering points of inflow connections, determining their legitimacy, assigning responsibility for correction of such conditions, establishing inflow control policies where none have been in effect, and instituting corrective policies and measures backed by investigative and enforcement procedures.

Control of I/I in future sewer construction and the search for and correction of excessive intrusion of extraneous waters to existing sewer systems is an essential part of sewer system management. A sewer system cannot be rehabilitated and then be expected never to develop additional points of infiltration or inflow. Regular preventive maintenance and rehabilitation programs must be instituted to control extraneous water flows. Efforts to achieve higher effluent quality through advanced degrees of treatment and funds dedicated to maintaining more rigid quality standards in public waters will be thwarted or rendered financially unsound if I/I is permitted to rob sewers of carrying capacities and treatment plants of their process performance capabilities.

To determine whether I/I is excessive, rough cost comparisons between transportation and treatment or elimination of I/I through corrective action are made. If I/I is excessive, the next phase should be the sewer system evaluation survey.

The SSES consists of the systematic examination of the sewer system to determine the specific location flow rate and rehabilitation costs of each I/I source. The survey is intended to translate overall findings of the analysis program to firm conclusions regarding the presence, location, and degree of I/I. It must also determine which I/I intrusions are excessive, according to criteria stated in the Clean Water Act and U.S. EPA rules and regulations. Definitive cost-effectiveness studies supported by the actual findings of the evaluations survey are used to estimate the amount of I/I that could be eliminated compared to the cost of expanded physical facilities.

This conversion of preliminary findings to more detailed evaluation facts must be based on a thorough diagnosis of sewer system conditions. Thus, the findings of the SSES must dictate the nature of corrective actions, their costs, the means by which I/I will be controlled, and the basis for treatment plant capacity design decisions. The evaluation survey will determine the extent of sewer rehabilitation in a rational sequence, though an accomplishment factor must be considered because the estimates on flow that would be cost effective to maintain may not be 100% accurate.

If regulatory agencies concur that the analysis stage clearly demonstrates that excessive I/I does not exist in the system, the evaluation phase should not be undertaken.

*P*ROBLEMS

Infiltration and inflow affects the operation of sewer systems and pumping, treatment, and overflow regulator facilities and adversely affects the urban environment and quality of water resources. Examples of the detrimental effects include reduced capacity required by present sanitary wastewater flow and future growth; premature need to construct relief sewer facilities; surcharging of sewers and flooding of streets and private properties; bypassing of raw wastewater; excessive wear on pumping station equipment; higher power costs, bypassing of flows, and adverse consequences to treatment efficiency; and diversion of flow from secondary–tertiary treatment stages or bypassing of volumes of untreated wastewater to receiving waters.

The term infiltration covers the volume of groundwater entering sewer systems from the soil through defective joints, broken or cracked pipes, improper connections, manhole walls, or other means. Inflow includes the result of deliberately planned or expediently devised direct connections of sources of extraneous water to sewer systems. These connections dispose of unwanted stormwater or other drainage water and wastes to a convenient drain conduit, including the deliberate or accidental draining of low-lying or flooded areas to sewer systems through manhole covers.

Regardless of the source of waters that enter sewers and affect their ability to provide sanitation and drainage, the net result is the same: reduction of conveyance and treatment capacities. It is difficult to distinguish whether groundwater has infiltrated affected sewer systems through subsurface defects or has entered sewers from direct connections.

*E*VALUATION TECHNIQUES

In sewer system evaluation, flow measurement is undertaken to define some variation of a certain flow component with time or to define peak or minimum flow conditions. Because flow measurement results may be used in design of capital improvements, it is critical to use equipment and methods that reduce inaccuracies during flow measurements. It is also important to recognize the limitations of data being collected. Flow measurement techniques are covered in Chapter 6.

FLOW COMPONENTS. Sanitary sewer system flow has three components: base flow, infiltration, and inflow. Separation and quantification of these components is a prime objective of flow metering.

Base flow can be determined in several ways with varying degrees of accuracy. Water consumption data adjusted for seasonal peaks, irrigation, unmetered connections, and water meter inaccuracies are often used. Also, minimum flow rates can be measured to estimate infiltration rates, which then can be subtracted from metered flow during dry weather conditions. Base flow can also be determined by evaluating per capita water consumption estimates for residences and commercial establishments upstream of the metering location.

Infiltration can be derived by subtracting base flow from total metered flow during dry weather or by compiling flow isolation measurements. Inflow is measured during wet weather conditions and is determined by subtracting base flow and infiltration from data recorded during wet weather conditions.

Peak Rate Determinations. Flow data for peak conditions are typically desired for each flow component. Base flow peaks can be obtained from recorded data by subtracting infiltration determined during dry weather conditions. Rates of infiltration should be determined if possible through correlation of groundwater flow data and past periods of high groundwater conditions.

Two peak rates of inflow should be established. The instantaneous peak inflow rate often is required to size pumping stations, interceptors, and other equipment that must handle wet-weather surges. The peak daily flow rate may be used to size equalization basins and other flow storage or settling devices. Attempts have been made to correlate peak inflow rates with rainfall intensity (Nogaj and Hollenbeck, 1981). If peak inflow rates are desired from gauging data, care must be taken to ensure proper equipment is in use if the monitoring site surcharges. All overflows or bypasses from the drainage area also must be monitored.

Annual Volume Determination. Determination of the annual volumes of base flow, infiltration, and inflow can be predicted by evaluation of historical data in conjunction with results from a flow-monitoring program. Base flow volume often is taken as some percentage of total potable water sales to allow for meter loss and sprinkling, and care must be taken to include institutions or residences that use private sources of water but discharge to the sewer system.

Infiltration volumes must be adjusted for seasonal fluctuations of groundwater levels, which can be done by comparing monthly water sales to wastewater flow records and developing weighted distribution to reflect typical seasonal infiltration rates. Inflow volume can be projected proportionately by using metered data and a known precipitation volume and by obtaining the annual total precipitation.

To determine wastewater flow parameters, it is necessary to obtain data on wastewater flow, rainfall, groundwater, and water consumption. To determine infiltration and inflow for an existing sewer subsystem, flow-metering equipment must be in place, permanently or temporarily. In most sewer system evaluation studies, temporary equipment provides most flow information required to evaluate wastewater flow variations within sewer subsystems. Permanent metering installations at treatment facilities and pumping stations may provide additional flow data.

Preinstallation Considerations. Proper planning for flow metering is a critical step for any sewer evaluation monitoring program. The first consideration is the use of the data. Which parameters must be defined and how accurately? What is the long-range effect on, for example, plant design or interceptor sizing?

The length of the metering period may have some bearing on equipment selection. Long-term metering may warrant the purchase of a portable flume or construction of a weir, whereas short-term metering may not. An item often overlooked is the method of data storage by the recorder. Several methods are available: circular and strip charts, magnetic tape, memory chips, and telemetering to a remote location. High technology data collection methods such as magnetic tape memory, chips, and telemetering allow faster data processing by electronic equipment. A drawback to some magnetic tape and solid-state data storage devices is that the person maintaining the meter often is unable to read the stored data in the field and make positive interpretations about how the meter is operating.

Circular charts are well suited for on-site cursory evaluation of data. All data can be viewed by an experienced technician in the field when the meter is maintained. A limitation when using charts is the need for manual data compilation, which can be time consuming and is subject to human error. Short duration metering programs frequently dictate the use of chart recorders unless electronic data storage devices can be read and feedback given to field technicians on a timely basis.

The effects of bypasses or overflows must be assessed. Such discharges may occur only during wet weather and often require monitoring to avoid possible miscalculation of flow volumes during sewer evaluation metering.

It is typically desirable to subdivide a sewer system to evaluate base flow, infiltration, and inflow in each subsystem. Without bypasses and relief sewers, subdivision of a sewer system usually is simple, based on direction of flow. Field work required for subsystem delineation should, where practical, be coordinated with flow meter reconnaissance inspection.

It is also important to recognize the limitation of temporary flow-monitoring data, as accuracy at temporary open channel sites is seldom better than 10%. Metering inaccuracies tend to multiply and overlap in large flow-monitoring projects that have many metering sites and require multiple subtractions and additions to define flow within individual districts. Moreover, accurate measurement of low flows from small areas often requires construction of costly weirs or other primary control devices. Where possible, use individual meters to provide flow data from specific areas to eliminate addition or subtraction of flow, significantly improving data reliability.

Flow-metering programs should be preceded by field reconnaissance to define the hydraulic characteristics of a proposed monitoring site (typically manholes) and select alternative sites if unfavorable conditions exist. A desirable open channel metering manhole has a well-defined regular channel, moderate flow velocity to maintain self cleaning, good access, but no change in grade or direction. Flow conditions will provide information needed to determine the best-suited equipment.

Field reconnaissance of metering sites should include determining accessibility and pipe shape and size, measuring sediment and flow depths, and sketching and describing manhole location. Assessing surcharge potential also is important. Consistent surcharging during wet weather may require the use of pairs of leveled depth recorders or flow velocity and depth-recording equipment because nonuniform flow or backflow conditions may occur.

Existing lift stations can also be used to monitor flow. Running times for constant-speed pumps can be recorded with strip-chart recorders or digital recorders and converted to meaningful flow data if good pump capacity information is available. For pump station monitoring reconnaissance, a determination of the number of pumps and their operational status is needed. Determining pump motor voltage may be required before certain event recorders are ordered. Ready access to the pumping station wet well may be needed to perform drawdown capacity tests.

Installation of primary control devices such as weirs or flumes generally will increase the accuracy of the flow-metering program, particularly under unfavorable hydraulic conditions. Weirs are probably the most accurate control section for low-flow conditions, though solids deposition and buildup of rags on or behind a weir installation can cause problems unless the backwater area is cleaned regularly.

Flumes present less of a maintenance problem but their applications must be limited to sites where manhole clearance is sufficient and excess approach velocities do not exist. Because control devices tend to reduce a sewer's hydraulic capacity, the installation of such structures should be coordinated with sewer system maintenance personnel.

A field reconnaissance for flow metering should provide the information to make equipment selections. Metering equipment can be leased or purchased. Also, a number of companies provide wastewater flow-metering services for sewer evaluation surveys.

Meter Maintenance. Temporary wastewater-metering locations are subject to a wide variety of pitfalls and should receive weekly minimum maintenance and inspection checks. A maintenance routine should include the manual measurement of liquid level and a cross check of measured value against recorded flow depth. Other maintenance procedures depend on the equipment used. It is advisable to perform velocity measurements with manual flow measurements at open channel sites to check equipment that automatically integrates area and velocity. A field log with notations for each visit to a site is recommended. Date, time, manually measured flow, and recorded flow should be logged with any velocity data.

Data Evaluation. Flow data evaluation, as required for a sewer evaluation survey project, typically involves producing flow parameters and hydrographs for each metering location. Data can be analyzed manually or automatically. Automated data analysis, like manual analysis, must be regularly checked for accuracy and validity. Proper quality assurance and quality control procedures should be developed and followed for both types of analyses. The benefits of applying computer technology to flow data analysis include increased accuracy and production from the use of peripherals such as plotters, printers, digitizers, and tape readers. Data processing has advantages over large projects, but manual analysis provides equally good results along with the opportunity to individually access special conditions that may not be decipherable by a computer.

Normally, data are analyzed on 15-, 30-, or 60-minute intervals. Flow rates for each interval period are determined and a total daily volume is computed. Peak rates can be determined from the hourly data.

Hydrographs are produced on a similar basis with each data point falling at the appropriate interval. Figure 4.1 shows a typical weekly hydrograph for a monitoring site, with hourly rainfall data superimposed.

PRECIPITATION MEASUREMENT. In sewer system evaluation, precipitation measurement correlates rainfall with flow-metering data to determine the amount of inflow entering the sanitary sewer system. Tipping-bucket or continuous-weighing rain gauges will provide information on rainfall intensity, total volume per event, and duration of the event, and charts are available for several durations. With some recorders, totalizers are available and provide a check against recorded data. In colder climates, snow-melting devices are also available on some models. For less sophisticated information, a manually read, graduated cylinder that records daily rainfall may be appropriate.

Before setting up a precipitation measurement system, other available data should be evaluated. Sources of precipitation data include the National Oceanic and Atmospheric Administration (NOAA), airports, state weather observers, electronic media weather observers, other public works and research agencies, and private citizens. The NOAA has an extensive nationwide network of recording rain gauges, and though not all sources will have rainfall intensity data, they may help determine the distribution of rainfall after an event. Those gauges with hourly rainfall data are summarized by state in a monthly publication entitled *Hourly Precipitation Data*. Another

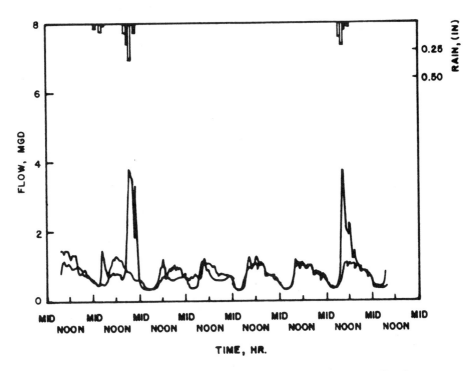

Figure 4.1 Typical weekly hydrograph for a monitoring site, South St. Paul, Minnesota (dry weather flow, 5/29/81 to 6/5/81; wet weather flow, 5/22/81 to 5/29/81; Mid = midnight; mgd × 43.83 = L/s; in. × 25.40 = mm).

useful publication containing daily precipitation quantities from all NOAA stations is *Climatological Data*, also published monthly for each state. Electronic media data from NOAA are also available for direct input to a computerized database.

GROUNDWATER GAUGING. Groundwater gauging locates the level of groundwater in soils and indicates the variation in this level. Two types of gauges commonly are used for sewer evaluation studies: the manhole gauge and the piezometer. The manhole gauge, shown in Figure 4.2, is used to determine groundwater levels adjacent to the manhole. The gauges are inexpensive and fairly easy to install, though they do clog easily from mineral deposits. The piezometer generally is installed in a hole excavated by a powered flight auger. Though more expensive, piezometers are more permanent, less prone to clogging, and with proper maintenance should last for years and provide higher quality data.

Piezometers should be situated away from underground utilities and far enough off streets to prevent damage from street maintenance equipment. When installed in parks or other areas that will be mowed, they should be

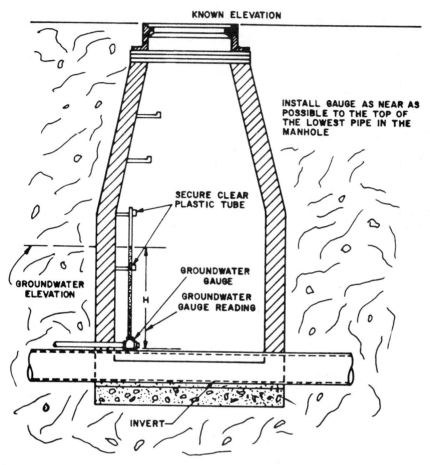

Figure 4.2 Static groundwater manhole gauge installation elevation.

staked or placed so that mower damage is unlikely. All caps should be threaded and locked if possible to prevent vandalism. Groundwater levels should be recorded weekly during flow-metering periods and biweekly or monthly thereafter. A plot of groundwater levels versus time is helpful in interpreting meter data and determining the levels of infiltration. In addition, certain field activities such as flow isolation measurements or television inspection can be scheduled with more confidence using groundwater elevation data.

SMOKE TESTING. Smoke testing is a relatively inexpensive and quick method of detecting I/I sources in sewer systems, and is best used to detect inflow such as storm sewer cross connections and point source inflow leaks in drainage paths or ponding areas, roof leaders, cellar, yard and area drains, fountain drains, abandoned building sewers, and faulty service connections.

Smoke testing also can be used to detect overflow points in sewer systems if groundwater is below the sewer.

Smoke testing should not be applied to sewer lines suspected of having sags or water traps, as such pipe conditions may prevent the smoke from passing through, causing false conclusions. Similarly, the method should not be applied to sewer sections carrying maximum flow. Smoke testing cannot be used to detect structural damages or leaking joints in buried sewers and service connections when the soil over pipes is saturated, frozen, or snow covered. In each case, the smoke will be prevented from escaping through the surrounding soil even though there are cracks or leaking joints in the pipes. Rainy and snowy days are not suitable for smoke testing, and testing should be monitored closely on windy days when smoke coming out of the ground may be blown away quickly and escape accurate detection.

Positive findings during smoke testing pinpoint I/I sources. Negative findings, however, do not necessarily prove that I/I does not exist.

Smoke bombs or canisters are used to generate the smoke required for the test. Smoke should be nontoxic, odorless, and nonstaining. Three- and five-minute bombs are typical, although there are bombs with both longer and shorter durations. An air blower forces smoke into the sewer pipes, and a gasoline-driven blower is most convenient for this purpose. The air blower should have a minimum capacity of approximately 1 500 L/s (53 cfs). A camera is used for documentation of smoke coming out of the ground, catch basins, pipes, or other sources during the test. Sand bags or plugs can be used to block sewer sections to prevent smoke from escaping through manholes and adjacent sewer pipes.

Police and fire departments should be notified daily of test locations, and residents may be informed of tests through the media. Residents should be informed individually on the day of testing, and personnel issuing notices and performing tests should have proper identification. A sample written smoke testing notice used is shown in Figure 4.3.

Procedure. Suggested chronological steps for smoke testing are as follows:

- Isolate line to be tested with plugging, up to 100 m (328 ft) at a time, noting any surcharged line sections. Smoke will not pass through a flooded section, which may require extra setups.
- Prepare a basic smoke sketch, including location, crew chief initials, and date.
- Commence smoke testing, using one blower at each manhole and enough bombs to ensure smoke travels throughout the test section. Larger diameter sewers may require more bombs or shorter test sections. Smoke should be generated continuously while visual inspection and photography are in progress.

ATTENTION
Smoke Testing & Sewer Survey

(City Name)

For the next few weeks, inspection crews will be conducting a physical survey of the _____ sanitary sewer system. This study will involve the opening and entering of manholes in the streets and easements. An important task of the survey will be the "SMOKE TESTING" of sewer lines to locate breaks and defects in the sewer system. The smoke that you see coming from the vent stacks on houses or holes in the ground is NON-TOXIC, HARMLESS, HAS NO ODOR, AND CREATES NO FIRE HAZARD. <u>The smoke should not enter your home unless you have defective plumbing or dried up drain traps.</u> If this occurs, you should consult your licensed plumber. In any event, if the harmless smoke can enter through faulty plumbing, the potential exists for dangerous sewer gases to enter your home. Should smoke enter your home, you may contact a member of the smoke testing crew working in the area and he will be pleased to check with you as to where and why the smoke has entered your home. If you have any seldom used drains, please pour water in the drain to fill the trap, which will prevent sewer gases or odors from entering the building.

Some sewer lines and manholes are located on the backyard easement property line. Whenever these lines require investigation, members of the inspection crews will need access to the easements for the sewer lines and manholes. Homeowners do not need to be home and the workmen will not need to enter your house.

Photographs are to be made of leaks occurring in the system. We anticipate the smoke testing will require approximately _____ weeks in your area. Your cooperation will be appreciated. The information gained from this study will be used to improve your sewer services and may reduce the eventual cost to taxpayers.

THANK YOU.

Name
Address
Phone No.

Figure 4.3 Sample smoke testing notice.

- Visually inspect the entire area by walking around front and back yards and around buildings. Watch for smoke leaks; typical sources are roof leaders, driveway drains, house foundations, holes in the ground over sewers or services, areas around manholes, and storm sewer inlets. Roof vents should not be considered as smoke leaks.
- Photograph all leaks, even if they do not show any water.
- Show the location of leaks on a sketch. Include the photograph number and directions taken and a description of the leak, including ad-

dress (or house number on the sketch); provide ties to the leak and area and type of surface drained by the leak.

Photographs. Photographs should show the maximum amount of smoke possible from the leak and the exact source of smoke, and should be taken from far enough back to provide a physical reference to the location of the smoke. Photographs of nothing but grass lawn are little help in finding the leak location at a later time; the same photographs showing some of the house would be better. Photographs should be numbered consecutively to ensure leaks can be identified at a later date.

MANHOLE AND PIPELINE VISUAL INSPECTION. Visual inspection of manholes and pipelines in the sewer system provides additional information concerning the accuracy of system mapping, the presence and degree of I/I problems, and the general physical condition of the system.

Manholes or entry points must be located before any underground inspections are conducted. Concise, up-to-date maps usually can be obtained from the owner's maintenance or engineering group. The accuracy of maps should be confirmed with operating personnel.

If accurate maps are unavailable, a preliminary working set must be prepared using all existing plans, reports, and record drawings; discussions with operating and engineering personnel familiar with the system are also important. Another source of data could be former operating personnel and engineers or long-term residents. The interviews must be directed to strategic areas of concern, such as the location of demolished buildings and the rerouting of streets. All information must be evaluated, and pertinent data should be used in the preparation of preliminary working maps.

After preliminary system mapping, field work can begin. From known locations, the system can be traced by opening manholes, identifying incoming and outgoing lines, establishing line-of-sight, and pacing off to the next manhole using distances on the maps. Building changes, street rerouting and paving, and landscaping can be complicating factors. Metal detectors, probe rods, transits, measuring wheels or tapes, dye tracing, and smoke testing can aid in solving many problems.

Starting in upper areas of drainage basins and following the flows from there typically is productive. Sewers near drainage basin upper boundaries generally are laid in much shallower trenches following the natural drainage of the ground surface. Conditions governing the installation of a pipeline help in locating that pipeline. The location of manholes should be recorded on the maps. If further work might be performed at a given location, it is useful to paint the manhole's identifying number on the cover.

Inspection. The actual inspection of manholes and pipelines depends on the type of information required. For instance, if the problem has been determined to be inflow only, a surface investigation may be all that is required to assess the conditions. However, if there is suspicion of groundwater leaking into pipes through deteriorating joints, it may be desirable to inspect the pipes visually by using a closed-circuit television camera (CCTV) or, if large enough, by actually walking the line. Inspections for the presence of infiltration should be performed during a period of high groundwater.

Data that can be obtained from the surface include

- Exact location of the manhole;
- Diameter of the clear opening of the manhole;
- Condition of the cover and frame, including defects that would allow inflow to enter the system;
- Whether cover is subject to ponding or surface runoff;
- The potential drainage area tributary to the defects;
- Type of material and condition of the chimney corbel and walls;
- Condition of steps and chimney and frame-chimney joint;
- Configuration of the incoming and outgoing lines (including drops); and
- Signs of frame-chimney leakage.

Additionally, the following data can be obtained by entering the manhole and using equipment such as portable lamps, mirrors, rulers, and probe rods:

- Type of material and condition of apron and trough;
- Any observed infiltration sources and the rate of infiltration;
- Indications of height of surcharge;
- Size and type of all incoming and outgoing lines; and
- Depth of flow indications of deposition and the characteristics of flow within all pipes.

By viewing incoming and outgoing lines with a mirror, it is possible to determine (for short distances into the lines) structural conditions, presence of roots, condition of joints, depth of debris in lines, depth and approximate velocity of flow, and location and estimated rate of any observed infiltration.

Safety Measures. When performing manhole or pipeline inspections, safety measures start when the vehicle arrives at the work site. It is important to observe the latest Occupational Safety and Health Administration confined space entry regulations. Many municipalities require police details or permits to be issued before allowing any diversion or disruption of traffic. If possible, work should be scheduled to avoid rush hour traffic, and signs and flagbearers must be placed far enough in front of the work area to warn motorists to slow down. Evenly spaced traffic cones should be used to channel traffic around the work area, and placing work vehicles between oncoming traffic and the workers is an effective shield in most cases (U.S. DOT, 1988).

Employees who are to enter a manhole should be in good health, should have no open sores or skin irritations, should not be under the influence of alcohol or drugs, and should have immunizations up to date. If a manhole that is unusually filthy or highly odorous must be entered, the walls and apron can be washed down with a high-velocity stream of clear water approximately 1 hour before entry.

There are six major categories of hazards encountered in entering manholes, presented in descending order of known frequency of accidents and deaths to sewer workers.

Safety precautions and procedures enumerated here do not take the place of a comprehensive safety training program and organizational procedures for entering manholes. Only properly trained and equipped personnel should enter or assist those entering manholes.

ATMOSPHERIC HAZARDS. Atmospheric hazards consist of three major types—explosive gases, toxic conditions, and oxygen deficiency. Unpleasant odors in sewers normally are only dangerous because they shift attention from other lethal conditions.

Explosive Gases. Explosive or flammable gases can develop at any time in a collection system. Methane is a product of biological decomposition prevalent in stagnant conditions caused by restricted flow. Even though methane is lighter than air, the danger is not limited to the arch and upper manholes of a system. Propane, gasoline, and many flammable solvents that find their way into the sewer are heavier than air and tend to form pockets in the lower reaches of the collection system. Care should be exercised in all sections of a line.

Toxic Conditions. Toxic conditions are likely a result of the formation of hydrogen sulfide, which is produced by the decomposition of materials containing sulfur. Hydrogen sulfide is heavier than air, accumulates in the lower sections of a collection system, and is easily detectable in low concentrations because it smells like rotten eggs. High concentrations of hydrogen sulfide, however, tend to deaden the olfactory senses.

Oxygen Deficiency. The oxygen concentration in a manhole may be reduced or even eliminated if the air is replaced by another heavier-than-air gas. The life expectancy of anyone entering a manhole with no breathable oxygen is approximately 180 seconds, and blinding can occur in less time.

PHYSICAL INJURY. There are several causes of physical injury during manhole inspection. The manipulation of tools in restricted spaces with unsure footing often results in bruises and strained muscles, and corroded steps can lead to falls and cuts. Tools being dropped to workers and then thrown back up and out again can cause eye injuries and facial cuts.

INFECTIONS. Infections are always a risk when entering a manhole, and there is the danger of infection from the unsanitary environment after any injury inside a manhole. Personal cleanliness is important, as every disease, parasite, and bacteria of a community can end up in the wastewater collection system.

ANIMALS. Insects and animals, although less dangerous to workers than infection and diseases, can be a hazard. Before they are entered, manholes should be inspected for insects, spiders, rodents, and snakes. Where insects are a problem, spraying with an insecticide is suggested. Periodic tests for Lyme's disease should be taken in areas known to be infested with deer ticks.

CHEMICALS. Exposure to toxic acids, bases, and other hazardous liquid or solid chemicals discharged to the wastewater collection system either by accidental spill or deliberate action is always a potential health hazard. Protective clothing is effective in guarding against these toxicants.

DROWNING. The chance of drowning while working in a manhole is increasing with the construction of more and bigger interceptor pipelines, and harnesses with lifelines attached should be worn when entering a manhole.
The following should be considered when entering a manhole:

- Never use only hands to open a manhole. The cover should be opened with proper tools and moved away from the opening.
- Manholes upstream and downstream should be opened to encourage natural ventilation, and a blower should be used to ventilate the manhole with fresh air. If the blower is gasoline powered, the exhaust must be downwind of the manhole.
- The area immediately around the manhole, including the ring, should be swept clean of any loose debris or dirt.
- A gas detector should be used the entire time a worker is in the hole.
- A person entering a manhole should wear a hard hat, gloves, steel-toed shoes, long-sleeved coveralls, and a safety harness.
- The lifeline should be connected to a man-rated, mechanical retrieval device and monitored by a crew member above ground, whose only job should be watching the person in the hole at all times. Tying the lifeline to an object is unacceptable, for if a passing vehicle were to hit the lifeline or the object to which it was tied, the worker in the hole could be seriously injured. After completing the task, the topside crew member should assist the exit of the manhole worker.

Data Recording. Data to be obtained from each inspection should be tailored to the objectives of the particular project. Checklist-oriented recording forms are more appropriate than those requiring a long narrative, though it is important that the forms include an area to record notes about conditions observed during the actual inspection. In addition to a written log, the inspection should be documented by photographs or videotapes and verbal descriptions of each defect. This procedure may be modified to record only defects as they are located and not the actual progress of the inspection, depending on documentation requirements.

BUILDING INSPECTION. A building inspection program will reveal sources of inflow located in the public sector that are not detected by way of other techniques. This process involves a systematic inspection of every structure, including basements, downspouts, and external areas that have potentially connected area drains, sump pumps, or other connections that might introduce stormwater to the sewer system.

Critical program start-up tasks include newspaper publicity, police department notification, letters of introduction, identification badges, building inspection crew orientation, and aerial map or lot map preparation.

Newspaper publicity will help to inform the community about the building inspection program. This will help to prevent incidents, keep refused entries to a minimum, and maintain good public relations. A public notice printed with a bold border should be placed in the local newspaper with the largest distribution.

When placing the public notice, the importance of the building inspection to newspaper readers should be explained to the newspaper staff. They will typically cooperate by moving the notice to a prominent location in the paper where more people are likely to see it. The public notice should appear in the newspaper the week before building inspections are scheduled to start.

It is also helpful to send a press release concerning the building inspection program to the newspaper the week before inspections begin. The press release should receive front page coverage if the newspaper is given enough lead time. It should be noted, however, that reference to "illegal sump pumps" should be avoided. It may be advantageous to issue another press release later in the project to remind people about the program and to maintain good public relations.

The police department should be notified about the building inspection program. The police should be informed of when the program is to start, the type of identification that inspectors are to carry, the names of the inspectors, at what time the inspections are to be performed, and any other aspects of the building inspection program that might help prevent incidents during the inspections.

A letter of introduction signed by the municipality or utility authority should be obtained several weeks before the inspections begin. Each building inspection team should carry a copy of the letter during inspection activities.

All building inspectors should be issued photoidentification badges that must be worn at all times during the inspection. These badges should be prepared before the first day of inspection.

Aerial or lot maps should be obtained, if available, from the municipality or utility authority. Manholes and connecting line segments should be marked before beginning building inspections.

All personnel conducting building inspections should be fully oriented to the program before they conduct inspections. The first day of work for the building inspection crews should be devoted to learning in detail how the form is to be filled out, the various sump types, dealing with refused entries, irate people, septic tanks, and vacant homes. The project manager should attempt to spend as much time as possible with the crews during the first few days of inspections. This should include a debriefing session at the end of the day. Personnel with previous building inspection experience should accompany new inspectors to the homes on the first day of the inspection if at all possible.

Building Inspection Program Procedure. Building inspections should be performed by two-person crews. Crews should begin the first inspection pass through the municipality in an orderly fashion. Each crew should be assigned a specific area or basin of the municipality and be responsible for all inspections in their area. Each crew should be given the appropriate aerial maps for their assigned area or basin. On the first pass, the crews should mark the street address on the aerial for each building where an inspection was attempted, write up a building inspection form, and submit aerial maps to the supervisor. Aerial maps should be marked to reflect actual conditions. This may require drawing in a new home or crossing out buildings that no longer exist.

At the end of each day on the first pass, each crew is to fill out the line segment number, street address and name (if available on mailbox) of the occupants for remaining uninspected homes on the crew's assigned street. This includes homes on both sides of the street and up to the corner (if the front door of the corner home faces the street). The streets that each crew has inspected or written forms for that day should be recorded on a master map. A master list of the number of inspections, refused entries, septic tanks, and vacant homes turned in each day for all crews should be maintained. All inspected, refused entry, septic tank, and vacant home forms should be collected each day. During the first pass all not-home forms are also collected daily. Not-home forms should be photocopied, filed, and the originals returned to crews.

Completed forms collected should be logged in on the weekly production summary and then reviewed for completeness of data the following day. It is important not to fall behind in form review and crews should help with this task as necessary. Reviewed forms would be filed by basin according to whether the status is inspected, refused entry, septic tank, or vacant home.

To ensure that all buildings have been accounted for, filed forms should be checked against aerial maps following first pass completion of an assigned area or basin. The number of buildings on the aerial maps should coincide with the building form count before that crew moves on to another basin or area. As crews perform inspections on second and third passes in assigned basins, corresponding field photocopies of not-home forms should be discarded and the completed forms should be reviewed and filed. Field forms following third pass completion of a basin are then ready for input to a database.

Commercial buildings can be inspected by one person if a building inspector is absent on a particular job. This task should be assigned to the most qualified inspector on the project.

Quality control of building inspections is especially important during the first few weeks of the program. Both internal and external checks should be done by a supervisor or another crew on at least three buildings per crew per week for the initial 2 weeks. Random internal checks should be done by a supervisor and a building inspector to reinspect sump pumps and sanitary sewer hookups. Other crews should do random external checks on buildings for address and physical building details. Further quality control checks performed throughout the program will ensure standardization of inspections and will verify the accuracy of collected data.

DYE-WATER TESTING. Dye-water testing is a rainfall simulation technique used to identify specific defects that can contribute I/I during rainfall or snowmelt. Additionally, dye-water testing can be effective in quantifying the amount of I/I that can enter a section of sewer or specific defect under a controlled runoff situation.

Equipment needed for dye-water testing is limited to that required to carry the water to the testing site and block the sewer or study areas before testing. When fire hydrants are close to the sewer sections to be tested, a fire hose is all that is needed to deliver water to the testing site. When the water source is not close by, water tankers are needed to deliver water. Sand bags or sewer pipe plugs normally are used to block sewer sections.

Fluorescent dyes typically are used, each having a distinct color readily detectable by eye. A suitable dye should be safe to handle, visible in low concentrations, miscible in water, inert to solids and debris in the sewers, and biodegradable.

Depending on the I/I sources to be identified and the configuration of the runoff situation being simulated (storm drain, drainage ditch, or spotflood), the procedures for dye-water testing differ. Five examples are provided below.

Determining Conditions Caused by Storm Sewer Sections. Storm drains that parallel or cross sanitary sewer sections and have crown elevations greater than the invert elevations of the sanitary sewers can be sources of rainfall-induced infiltration or inflow. They are inflow sources if there are cross connections between storm drain sections and sanitary sewers; they are infiltration sources if stormwater can exfiltrate from them, percolate through the soil, and enter the sanitary sewers through pipe or joint defects. General procedures for dye-water testing in storm drain sections are as follows:

1. Plug both ends of the storm drain section to be tested with sand bags or other materials and block all overflow and bypass points in the sewer section. Bypass flow around the section under test, if necessary.
2. Fill the storm drain section and stormwater inlets or catch basins to just below the grate with water. Add dye to the water.
3. Monitor the next downstream manhole in the sanitary sewer system for evidence of dyed water.
4. Measure the flow in the manhole before and during dye-water testing. As an alternative, flow can be simultaneously measured at both the upstream and downstream manholes during the test.
5. Record the location of storm drains and sanitary sewer lines being tested; the time and duration of tests; the manholes where the flows are monitored and the flow rates; the observed presence, concentration, and travel time of the dyed water to flow-monitoring manholes; and the soil characteristics.

Determining Conditions Caused by Stream Sections. To determine whether stream sections, ditch sections, or ponding areas located near or above sanitary sewer sections are causing I/I, a procedure similar to that described above should be followed. Stream sections, ditch sections, or pond areas to be tested should be plugged or dammed (if necessary) and filled to desired levels with dyed water. Steps 3 through 5 above are then followed.

Identifying Sources on Private Property. Roof leader, cellar, yard and area drains, abandoned building sewers, faulty connections, and illegal connections are usually located on private property, and owners should be notified before any tests are conducted. To identify inflow sources, dyed water is poured into the expected source and looked for in the closest downstream manhole in the sanitary sewer system. The date of the test, the address where the inflow sources are identified, and the type of inflow sources should all be recorded. Again, weir or depth and velocity measurements can be made at upstream and downstream manholes to quantify the source.

Identifying Structurally Damaged Manholes. The dye-water test can also be used to identify structurally damaged manholes creating potential I/I problems. This is accomplished by flooding the area close to suspected manholes with dyed water and checking for entry of dyed water at the frame-chimney area, cone/corbel, and walls of the manhole.

Verifying Sources Found by Other Means. The dye-water test can verify suspected sources of I/I identified in a physical survey or smoke-testing study, and quantification of defects can be done at the same time. The log sheet from the field study is used to identify and locate the suspected I/I source if the area is flooded. In some cases, it may be necessary to restrict runoff with sand bags to allow the area to become saturated. The downstream manhole is monitored for presence of the dyed water. If a positive result occurs, a weir or depth and velocity measurement is taken to quantify the source. This works well for findings such as manholes subject to surface runoff, holes in the ground smoking over services or main lines, large areas of ground smoking over or near services or main lines, and cracks in the street pavement that are smoking.

Safety Measures. Dye-water testing is limited to locations where large quantities of water are available. It is usually not practical to flood an area that is more than 150 m (500 ft) from a fire hydrant because of the amount of equipment necessary and setup time. Tank trucks can be used, though they are limited in both the rate and total quantity of water that can be applied, and there may be considerable time lost in refilling the truck.

Both sanitary sewer and storm drain manholes may have to be entered for the test. Water head buildup behind plugs is dangerous, and safety is extremely important. In some places, water use may be restricted by either rate of use or total quantity available.

When a positive dye transfer has occurred on a line segment, it may not be possible to conduct another test on either that line or a downstream line for some time because of the presence of the dye. This is most limiting when more than one spot check is desired on a line segment. Delays can be limited under these circumstances by conducting spot checks, beginning with the most downstream line segment to be checked and moving progressively upstream.

When a positive dye transference is observed, it often is necessary to inspect the line internally so that the actual location of the leak and proper rehabilitation technique can be determined. It may be desirable to reflood the setup at the time of televising to identify the defect positively.

There are several important safety measures that should be observed when dye-water testing:

- A gate valve should always be used to control the flow of water from a hydrant, and hydrants should always be opened and closed slowly and completely. Never use the hydrant to throttle the flow because this may damage the hydrant. Use a pressure gauge when inflating plugs or bags, and never overinflate a bag or plug; damage to the pipe may occur or it may explode and injure the technician.

- Always attach a valved air line to a bag or plug so that it may be deflated from the surface. Water head buildup behind a bag or plug in a sewer line is powerful and dangerous, and if a bag or plug falls, the technician may be seriously injured. He or she may also be pinned against the manhole wall and possibly drown if enough water is backed up. The larger plugs present the most dangerous situations.
- Always watch the water level and be aware of the minimum level that will cause flow to backup to buildings and cause property damage.
- Provide proper traffic control where hoses cross streets. Damage to both the water system and cars can result from vehicles hitting hoses at high speeds.
- Always secure the discharge end of the hose. High pressure and flow can cause the end to go into a wild whipping action, injuring personnel and damaging property.
- Remove all plugs when a setup is complete. Failure to do this will result in backup and property damage.

Logging Test Results. A field log sheet should be filled out for each dye-water test, regardless of whether a positive transference is observed. Data that should be included on the log sheet include the date, time, field crew, location, type of setup, sketch of the setup, sanitary manholes checked, dye transference information, flow readings before and during transference, flooding time, pipe size and storm footage involved, and comments on the setup.

The sketch should indicate exactly what was flooded and the relationship between that and the sanitary sewer system. It is often desirable to photograph the setup, and the photograph number and direction should be recorded on the sketch for reference.

Observations such as previous and existing weather conditions, soil conditions, access for future maintenance or rehabilitation, unusual conditions in storm or sanitary systems, and difficulties incurred in performing the test can also be recorded on the log sheet.

NIGHT FLOW ISOLATION. Night flow isolation is a technique used to determine the amount of extraneous water, usually infiltration, entering a reach of sewer. It differs from other flow measurements in that the primary purpose of performing night flow isolation is to determine the specific reaches of sewer that have excessive infiltration, so that further action, usually internal inspection, may be performed.

For economical and reliable planning and execution of a flow isolation program, an accurate system map is needed. Proper selection of measurement points is critical to the success of an isolation program, and the map must accurately reflect the layout of the system. Efficiency of field execution often is affected by map reliability. The map should contain information on all pipe diameters and lengths and permit ready location of the manholes.

Additional information beneficial to performing night flow isolation includes manhole accessibility, estimate of flow rates, cleanliness of pipes, sizes of weirs and plugs required, large users, and night users.

The ultimate usefulness, accuracy, and cost of flow isolation depends on the selection of the proper length of reach to be isolated. The two major determining factors are measurement accuracy versus reach length and the effects of possible groundwater migration.

Regardless of the flow measurement technique used, the measurement will always have some component of error, which typically is in proportion to the flow rate being measured. Sources of these errors include both inherent limitations of the techniques and limitations that arise in practical applications. For example, an inherent limitation of a weir is the accuracy of the assigned coefficient of discharge that enters into the weir calibration. A limitation that arises in practical application is the accuracy with which it is possible to determine the water level over the notch while reading the weir in the field. Each measurement technique has its own sources of error. In general, an inaccuracy of up to 10% of flow rate can be expected for a properly executed flow measurement.

If the measured flow during flow isolation is purely infiltration, this error is not significant. Difficulties arise, however, when noninfiltration flow also is present. This is more evident when using a differential of two measurements to obtain a net measurement, as in differential isolation.

Two alternatives are available to reduce this type of problem:

- Reduce the total flow rate to be measured. This can generally be done only by plugging and is useful for moderate to small flow rates.
- Increase the amount of infiltration to be measured. This can be done by increasing the net length of reach under measurement. For this to be effective, the sanitary component of the flow rate generated within the reach must be reduced.

For both alternatives, the sanitary flow component of the measured flow should be as small as possible. For this reason, such measurements generally are conducted during periods of minimum flow, typically from midnight to 6 a.m. Differential measurement techniques generally are not suitable for infiltration isolation of short reaches.

Groundwater Migration. The selection of reach length to be isolated depends on the possible effects of groundwater migration. In most circumstances, infiltration to a pipe is not confined to singular sources but is distributed, with many potential entry points for any given groundwater condition.

Because groundwater excluded from one section of pipe often migrates along the pipe trench and enters at other defects, it is impractical to consider rehabilitation of defects on a point-by-point basis. If any rehabilitation of a pipe is attempted, all probable points of entry should be considered in the I/I correction and cost estimate.

Normally, isolation of infiltration down to small reaches is unnecessary. The length of reach to be isolated should reflect the length thought to have a common infiltration problem and should also reflect the length of reach that will be considered for rehabilitation as a unit.

Other considerations in the selection of measurement points and reach lengths to be isolated are access, suitability for measurement, the costs of obtaining the measurements, and the potential costs of internal inspection and rehabilitation. Selecting longer lengths of reach will result in fewer measurements and a less costly isolation program.

The selection of longer reach lengths, however, also will result in higher total costs for internal inspection and rehabilitation. The latter is somewhat mitigated by the fact that the cost per unit length of internal inspection and rehabilitation is lower for longer continuous runs of pipe than it is for the same length of shorter, discontinuous runs because of money saved in access and setup costs to perform the work.

Single-Pass Method. This method of isolation consists of prior selection of all measurement points from map studies and then execution of all field measurements. Its advantage is simplicity of planning and execution. The disadvantage of this method is that it is inflexible, thus planned measurement points may be unsuitable or unnecessary.

Multipass Method. This method of isolation involves selecting a few key measurement points for field execution and then analyzing the results before the selection of further measurement points. The multipass method is more complex but allows concentration of effort in high infiltration areas.

Circumstances should dictate the choice of single or multipass method. If infiltration is distributed, the single pass method is more useful and economical. If infiltration is confined to smaller areas, the multipass method may be less costly.

Isolation Techniques. The objective of the flow isolation program is to isolate small reaches of sewer and measure the infiltration rate within each. Two methods are available to isolate specific lengths of sewer: plugging and differential isolation.

PLUGGING. Plugging consists of physically isolating the sewer length from the rest of the system by means of plugs inserted to the sewer pipes. Plugs generally are pneumatic and available from several manufacturers in diameters up to 610 mm (24 in.) for rigid body plugs and may be available up to 1 067 mm (42 in.) in diameter for nonrigid bag plugs.

Because plugs need to be inflated, compressed air must be available at the site. Installation of larger diameter plugs may require winching. Plugs generally are inserted to an incoming line for easier removal.

Plugs may be subject to large total hydraulic and pneumatic forces even at low heads and small diameters. Failure of plugs is not uncommon. Workers near or downstream of plugs must exercise extreme caution at all times. All plugs should be tied off to manhole steps and have a stout tag line carried to the manhole entrance. The tag line is used to recover blown plugs and remove deflated plugs.

Plug removal is the most critical part of the plugging operation. Deflation is best done from ground level by use of an extension hose on the plug valve. If the manhole must be entered to deflate a plug, the person entering should wear a safety harness and should never stand in front of the plug, which should be deflated gradually. Even with care, a plug may be ejected from the pipe with high velocity, causing the manhole to fill rapidly with the released water. The person in the manhole should be positioned to avoid the plug and exit the manhole rapidly.

After a section of sewer is isolated by plugging, sufficient time must be allowed for the pipe to drain down before any measurements are attempted. This time should increase as length of isolated sewer increases and the average slope of the sewer decreases. The drain downtime may be estimated roughly by assuming a flow velocity of approximately 0.15 m/s (0.5 ft/sec) from the plug to the point of downstream measurement. Before any measurements are taken, however, an equilibrium flow situation should be verified at the point of measurement.

Plugs inserted in the pipe will interrupt the existing flow and cause wastewater to be stored behind the plug. Before plugging, the flow rate, volume of storage available, and elevation of the pipe relative to the lowest floors served should be estimated to determine the safe length of time the plug may be left in the line. The relationship of the safe plugging time to the drain down time of the pipe will determine the maximum length of sewer feasible to isolate by plugging. The estimated safe plug time alone should never be relied on in efforts to avoid flooding basements. Frequent visual observation of actual water depths in upstream manholes is vital during a plugging operation.

When a plug is removed, a surge of water may be propagated for downstream. This may interfere with downstream measurements, blow downstream plugs that are lightly seated, or flood lowlying downstream basements. For these reasons, plug removal should be as gradual as possible and properly sequenced with other operations conducted within the same sewer. The manufacturer's instructions on the proper use and maintenance of plugs should always be followed to avoid injury or property damage.

The advantages of plugging for flow isolation are physical isolation of the sewer under consideration and reduction of the flow rate required to be measured. This tends to increase the accuracy of flow measurements, especially if obtained by use of portable weirs.

DIFFERENTIAL ISOLATION. Differential isolation requires subtracting all flows coming into the section from all flows going out of the section to obtain the net increase in flow within the section. Flows are not physically interrupted, eliminating problems of upstream storage associated with plugged flows.

To obtain a usable net flow rate with this method, care must be taken to ensure upstream measurements do not have an adverse effect on downstream measurements. The entire system of flows and measuring devices associated with a given section of sewer should be in equilibrium, which is most important when using portable weirs for flow isolation because a certain amount of time is needed to stabilize them after installation. This can be achieved by setting all weirs and allowing the entire system to reach equilibrium before taking any readings.

Portable Weir Methods. A common measurement technique used in flow isolation is to insert a portable V-notch or other weir to the incoming pipe at a manhole. Portable weirs are commercially available to fit circular pipes. Advantages of portable weirs are relative ease of installation and their general calibration for direct reading of flow rates. For a usable measurement, care must be taken to ensure that

- The weir is installed level;
- The weir is properly seated and watertight at its perimeter;
- The nappe is aerated;
- The velocity of approach is small;
- The flow over the weir has reached an equilibrium condition; and
- The weir is properly read.

Under ideal conditions, a V-notch weir may have an error proportionate to the flow rate plus an error resulting from flow depth estimation, which increases with low flow rates. No real sewer system provides measurement points meeting all ideal conditions, so care should be taken to reduce adverse conditions. Portable weir measurements must be considered, at best, an estimate of actual flow rates and reasonable judgment must be used in interpreting results.

Despite these limitations, in most circumstances portable weirs are considered a practical and cost-effective method of obtaining flow isolation measurements.

Velocity-Area Method. Estimates of flow rates at a measurement point may be made using the velocity-area relationship:

$$Q = AV \qquad (4.1)$$

Where

Q = flow, L/s (cfs);
A = cross-sectional area, m^2 (sq ft); and
V = mean velocity, m/s (ft/sec).

To use this relationship, both the cross-sectional area of flow A and the mean velocity V must be determined.

The cross-sectional area of flow can be obtained by first measuring the depth of flow at the point where the velocity reading is obtained. The depth of flow may be measured by pipe caliper or, less satisfactorily, by ruler.

If sediment is present, the depth of debris must be estimated by seating a ruler on the pipe invert and marking the apparent top of the debris on the ruler by feel, withdrawing it from the flow to obtain the reading. The cross-sectional area of flow A may be computed by reference to prepared tables or by geometry. The sediment area should be subtracted from the total wetted area to obtain the actual area of flow.

The mean velocity V of the flow may be obtained by using magnetic or propeller current meters, dye interval timing, or timing of a floating object. The use of magnetic or propeller current meters to measure open channel flow in pipes is described in the literature, and the manufacturer's directions should be consulted for each particular type of instrument.

Dye interval timing is a reasonably accurate method of obtaining results provided that the following conditions are met.

First, the velocity of flow should be approximately 0.6 to 1.2 m/s (2 to 4 ft/sec). Slow flows may cause irregular dispersion of the dye slug or make determination of the boundaries of the slug difficult; extremely fast flows may cause difficulty in the timing of the leading and trailing boundaries of the slug and make insertion of a well-defined slug of dye difficult.

Also, the pipe should be clean, free of roots, uniformly graded, and the flow depth constant. Deviations in any of these factors will introduce inaccuracies to the technique. Without extensive examination, it may be impossible to determine if all such conditions are suitable for a given pipe. Furthermore, at the low-flow depths typical of night flow isolation measurements, the effects of these factors are magnified.

Dye interval timing may be useful in flow isolation under certain circumstances; however, because of the above limitations, it cannot be considered a generally applicable technique.

The use of floating objects to estimate mean velocity is subject to many variables and is not accurate enough to be of value in flow isolation. The most important item in conducting flow isolation is a reasonably accurate map of the sewer system, showing pipe and manhole locations, lengths, and diameters. Without such a map, it is difficult to properly plan or execute night flow isolation or interpret the results in a meaningful way.

Fluorometric Methods. Fluorometry, also known as dye dilution, is a useful flow measurement technique in sanitary sewers because it is independent of sewer dimensions, velocities, conditions, and surcharging. It is particularly useful in large-diameter sewers in which dye is continuously injected at a constant rate far enough upstream from the site of measurement that it is thoroughly mixed and its concentration is uniform throughout the cross section of wastewater at the point of measurement. Under these conditions, a rate of flow can be calculated anywhere downstream where dye is present by the following relationship:

$$Q_1 C_1 = Q_2 C_2 \qquad (4.2)$$

Where

Q_1 = the rate at which the dye is injected, L/s (cfs);
Q_2 = the discharge rate sought downstream, L/s (cfs);
C_1 = the concentration of the injected dye, mg/L; and
C_2 = the concentration of the dye at the point of measurement, mg/L.

The dye's change in concentration at any point in the sewer is proportional to the change of inflow rate. The proportionality is defined by the variables of the above equation. Of the four variables, three are known and are used to calculate the fourth. The flow rate or discharge rate sought downstream, Q_2, is calculated by

$$Q_2 = (Q_1 \times C_1)/C_2 \qquad (4.3)$$

Results. Flow isolation data are generally expressed in terms of (L/d/mm diam)/km [(gpd/in. diam)/mile]. The measured flow is divided by the pipe diameter and the isolated length.

Care must be used in drawing conclusions from the data. A high infiltration rate is not bad in and of itself. The rate must be evaluated both qualitatively and quantitatively. Other factors, such as sewer capacity, projected future needs, treatment costs, and basement or low-lying area flooding problems, must be considered. For example, many communities or sanitation districts impose strict requirements because of limited sewer capacity. Reducing infiltration increases sewer capacity available for growth. Conversely, cities with adequate capacity and low treatment costs can justify higher allowable levels of infiltration as cost effective. The accuracy of measurements is also a limiting factor.

Pipeline Cleaning and Television Inspection. Pipeline or sewer cleaning is necessary for both efficient collection systems operation and improving the effectiveness of television inspection. Standards for cleaning that are acceptable for system operation will provide for adequate television inspection results. However, standard system operation cleaning should be regarded as a minimum acceptable effort with regard to television inspection.

As with any cleaning, all the collected sediment and debris should be removed from the line and disposed at an approved site. Care should be taken during cleaning to ensure that only minimal amounts of deposition are lost to downstream lines. An outline of sewer cleaning procedures can be found in *Wastewater Collection Systems Management* (WEF, 1992).

Television inspection is accomplished by using closed-circuit systems specifically designed for sewer inspection. There are several configurations of CCTV systems for sewer inspections, each of which have the following in common:

- Power for operation generated on site;
- Power control;
- Transport winches;
- Video (color, if possible) and lighting control;
- Recording and documentation; and
- Radio communication.

Each equipment format has unique advantages and disadvantages; however, systems that allow the television van operator to control both the speed and travel of the camera and allow for remote control of the camera itself are the most efficient and provide the highest quality pictures. Additionally, systems that use only one or two conductors to control all viewing functions are the most reliable and the easiest to maintain.

The effectiveness of television inspection is directly related to the completeness and accuracy of the collected data. For each pipeline or sewer inspected, records should be collected on a field form, videotape, photographs, or a combination thereof.

The inspection form should contain the following data for each manhole-to-manhole section inspected:

- The date of the inspection;
- The reason for the inspection;
- The location of the pipeline and upstream and downstream manholes;
- The compass direction of the viewing and the direction of the camera's travel;

- The pipe size, type, pipe joint length, and overall footage of the inspected sewer;
- The quantity of I/I expected to be observed, the actual quantity of I/I observed, and the total quantity of I/I measured in the sewer at the time of the inspection;
- A description of each service connection and defect observed and its distance from the point at which the viewing began;
- The observer's assessment of rehabilitation recommendations for the inspected sewer; and
- Reference on the log to each photograph taken and the videotape of the entire inspection.

For each pipeline inspected, the videotape should contain the entire pipeline, regardless of its condition, to ensure the work was completed efficiently. A proper videotape permits a review of recommendations, the accuracy of the work, overall effectiveness of the entire television program, and provides a final check to ensure no defects have been overlooked.

The videotape also should include a brief and informative verbal description of the pipeline being inspected. The narration should contain no reference to cause and effect in regard to defects. Finally, the narration should contain no reference to any party's liability as it relates to defects, payments for rehabilitation, or personal opinion.

Photographs. Photographs should be taken of each severe structural defect and all significant sources of I/I. Typical pictures of the pipe or the pipe joints are optional. Photographs are acceptable when there is an urgent need to illustrate defects to interested parties.

Leak Quantification. The quantification of I/I sources during a sewer system evaluation survey is based on the type of source and physical characteristics of the sewer system or the area of the leak. In many cases, they are based on empirical values instead of engineering principles. Experience is a valuable tool both for determining the quantity of a source and obtaining proper physical observations to make a good estimate of how much I/I is leaking to the sewer system.

To determine the quantity of a leak, it is necessary to know whether the source is infiltration or inflow. They can be distinguished by the duration of their response to rainfall. Regardless of how leaks enter the system, those that last from a few days to several months are considered infiltration. Those flows that last a much shorter period are considered inflow. Inflow can be further divided into direct and indirect sources. Direct inflow sources enter the system directly as a result of runoff and have a short duration. Examples of direct inflow sources are catch basins, roof leaders, and manhole covers.

An indirect inflow source is one in which runoff has to pass through a medium before it enters the sewer system. Sources of this nature include pipe joint leaks from transfer of water from storm drains or areas of surface drainage; manhole wall leaks from the same conditions as the joint leaks; frame-chimney joint leaks; and foundation drain leaks.

Similar to infiltration sources, these sources have a longer duration and may contribute a more constant flow throughout a rain event. Direct inflow sources may contribute a larger rate of water but only for a short period of time. Almost all direct inflow sources can be quantified by application of the rational formula; with a trained field technician obtaining physical measurements, good estimates can be determined.

Indirect inflow sources are largely discovered during dye-water testing. The most accurate method of determining the quantity of leaks in the pipe is to install a portable weir to the line before dye-water testing. After transference, reinstall the weir again to measure the increase in flow caused by the line defects in the system. During subsequent television inspection, the previously measured increase in flow will give a basis for the quantity of flow from these defects.

In some cases, it is not always possible to obtain a weir measurement because of the condition of the line, and it is only possible to obtain a depth measurement both before and after the increase in flow. To estimate this quantity, Manning's equation for open channel flow could be used:

$$Q = (k/n)R^{0.67}S^{0.5}A \tag{4.4}$$

Where

Q	=	the flow rate, L/s (cfs);
n	=	the Manning (Kutter) roughness factor, dimensionless;
R	=	the hydraulic radius, m (ft);
S	=	the energy loss per unit length of pipe, m/m (ft/ft);
A	=	the area of the discharge, m^2 (sq ft); and
k	=	a conversion factor: 1.00 for metric and 1.486 for English units.

Probably the most difficult sources to quantify with any degree of accuracy are manhole lids subject to surface runoff. The amount of flow entering through lids is based on the ability of a variety of manhole lids to contribute inflow under different heads.

When not enough data are available to allow the use of flow equations, such as with leaking manholes and small leaking joints, estimates must be made at the time of discovery based on prior training and experience in judging sources. In some cases, leaks can be captured in a bucket of a known volume and timed as to how long it takes to fill, a simple but effective method.

The accuracy of all I/I estimates is influenced by the empirical nature of the task. Study conditions do not always match the parameters of the flow equations used; in these cases, engineering judgments are needed to apply the equations. These judgments are based on past experience in quantifying I/I sources and knowledge of the sewer involved in the study.

EVALUATION OF INFILTRATION

There are various infiltration quantification techniques to estimate infiltration entering a sewer system. Because the initial interest with infiltration quantification is with the entire sewer system, these techniques are more applicable to the entire system than to a particular subsystem. The following techniques can be used to estimate total infiltration to a sewer system, and the available information for the system will influence the method used.

Some of these methods will use data that, in many instances, will include inflow and infiltration. If inflow is a significant portion of the total extraneous water gaining access to the sewer system, additional estimates will need to be made for inflow, and these quantities will need to be subtracted from the totals obtained by use of the methods described here. In many systems, where infiltration is the prime source of extraneous water, inflow makes up only a small fraction of the total volume of extraneous water gaining access to the system.

WATER USE EVALUATION. The water use evaluation method makes use of water supply records for the purpose of estimating the amount of domestic wastewater discharged to the sanitary sewer system. This requires obtaining monthly water use records, and an estimate of the portion of the water that is reaching the sanitary sewer could range from 70% in summer months to 90% in winter months. This gives an estimate of domestic, industrial, and commercial wastewater flow rates. With this information, flow rates can be subtracted from the total flow measured at the area's wastewater treatment facility to obtain an estimate of infiltration entering the sewer system.

The amount of water that eventually will reach the sewer system depends on the amount of outside-of-house water use. The amount of water used for lawn and garden irrigation must be estimated. Some of the water applied to lawns and gardens could reach the sanitary sewer system, but this is difficult to estimate and is neglected.

When working with water records, water not accounted for is another factor that needs to be considered. Unaccounted water is the difference between the total water supplied to the water system through wells, springs, or reservoirs and the amount of water measured from each of the individual water meters on each user's water connection. Although water distribution system leakage is a major reason for water not accounted for, there are other causes, such as incorrect or inaccurate meter reading because of inaccuracies of either the meters measuring the input to the system or the individual user's meter. In the latter case, improperly measured water could still be directly entering the sewer system, whereas in the case of leakage, water would not be entering the sewers directly. Illegal taps and unmetered withdrawals from firefighting lines, street-flushing fire lines, and hydrants are other sources of unaccounted water. Water from illegal taps could easily enter the sewers, although water from unmetered use of fire lines and hydrants may or may not.

All of these factors will have a bearing on the estimate of the amount of water that eventually becomes wastewater, and they should be considered in each individual case before estimates are made. If an area is supplied with a secondary water system, eliminating the need to use water for outside watering, the amount of water reaching the sanitary sewer system will be a large portion of the total water supplied to the system. This portion could be in the 85 to 90% range.

Maximum–Minimum Daily Flow. The theory behind this method is that infiltration is constant throughout any given day. If there is no precipitation, then the daily flow increase is strictly attributable to the domestic flow contribution. Industrial flows are also assumed to be constant throughout the day. If this principle is applied to an entire month, infiltration can be assumed to be constant and the variation between the maximum and minimum daily flows is assumed to be the domestic flow rate.

The above principle can be applied to treatment plant influent flow data to estimate the amount of domestic flow. To determine the estimate for infiltration, the domestic flow is subtracted from the total flow at the treatment plant and the industrial flow estimate also is subtracted. This procedure can be carried out with monthly averages to obtain the estimated infiltration for the entire year.

Maximum Daily Flow Comparison. This method is based on the principle that, if infiltration does exist in the sanitary sewer system, the influent flow rate measured at the treatment plant cannot drop any lower than the rate of infiltration. This suggests that, if the entire population of the area being considered were sleeping, all the industries were shut down, no commercial functions were active, and no precipitation was present, the flow to the treatment plant would be entirely infiltration because there would be no domestic, industrial, or commercial contributions. For the sake of simplicity, this method can be assumed to occur during the early morning hours in the service area. That infiltration remains constant throughout any given month or week, depending on available data, also should be assumed.

A graph should be made of the minimum daily flow at the treatment plant for each month during the period of interest, bearing in mind that the quantity plotted should not be the average minimum flow for the month, but the lowest minimum daily flow seen during that month. If these points are plotted for each week or month during the year, the difference between the maximum and minimum points on this curve can be assumed to be the annual infiltration for that year. The maximum point on the curve can be considered to be the high groundwater infiltration rate and the minimum point can be considered the low groundwater infiltration rate. A good check would be to verify that low infiltration does coincide with dry periods.

The drawback of this method is that it does not properly consider low groundwater infiltration, which could represent a significant portion of the total annual infiltration. Therefore, the closer low groundwater infiltration rate is to zero, the more accurate this estimating method will be.

Nighttime Domestic Flow Evaluation. This estimating procedure is based on a method of flow prediction outlining relationships of maximum and minimum daily flows with average daily flows based on population (WPCF, 1969). These relationships are for domestic wastewater flow only and are considered to be a fairly reliable tool for estimating.

Figure 4.4 represents a typical dry weather day and shows how the various components of the total wastewater flow are related. The graph reflects the assumption that infiltration, industrial, and nighttime domestic flows (NDF) remain constant throughout any given day. The NDF is defined as that portion of the total minimum daily flow attributed to domestic activities.

Small Subsystems. The best and most accurate method of estimating infiltration from small subsystems is by direct measurement techniques described earlier in this chapter. Some of the estimating methods described here for the total system have limitations when applied to small subsystems.

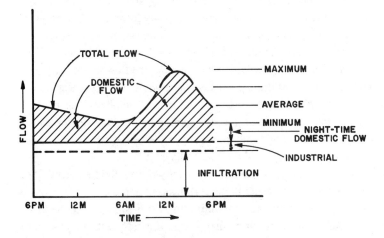

Figure 4.4 Daily component flow comparison.

For instance, water use data are generally not readily available for small subsystems. Also, many of the assumptions made with respect to outside watering characteristics become more crucial to the accuracy of estimates for small subsystems.

The other three methods discussed under total system techniques can be used in small subsystems, all requiring that direct flow measurements be taken. If direct flow measurements are taken, however, these methods or variations of these methods can be used to estimate infiltration from the area of interest.

When infiltration flow rates or estimates are obtained, unit infiltration or incremental infiltration rates should be calculated. Incremental infiltration is defined as the infiltration flow rate (L/d/mm diam)/km ([gpd/in. diam]/mile) relative to system size. The calculation of incremental infiltration requires information on the characteristics of the sewer system, and the total length of various size pipes within the area of interest must be obtained. Incremental infiltration rates vary widely from city to city and region to region. An incremental infiltration rate may be high for one area but low for another area of the country. The significance of the incremental infiltration calculated for any particular sewer system will depend on the characteristics of the area in which the sewer system is located.

INFILTRATION LOCATION. The methods described above help to obtain quantitative estimates of the amount of infiltration in the system. Further flow measurements at key locations within the system can help pinpoint infiltration. Other methods of confirming the results of flow-monitoring techniques are available and can also be used to locate infiltration areas within a system when flow-monitoring methods or information is not available.

One method of estimating locations of infiltration is to measure the temperature of the wastewater within the system. This can be done during a manhole inspection, typically a part of a sewer system evaluation, and wastewater can be measured throughout an entire area or system. Temperature data can provide infiltration information if the temperature of the groundwater can be established.

Groundwater temperature can be affected by several factors: the temperature and location of the source of the groundwater, the air temperature and location of the source of groundwater, the type of material transporting the groundwater, and the distance groundwater must travel. The temperature of the groundwater source, in many cases, is a nearby river, stream, creek or large body of water, which can be obtained from local 303 (e) studies, 208 Area Wide Water Quality Studies, U.S. Geological Survey reports, and Soil Conservation Service reports. These documents can also be used to obtain soil information in the area of the sewer system.

Groundwater temperature within the area of a sewer system can be compared with the wastewater temperatures taken during a manhole inspection to project the possible existence of infiltration within an area of interest.

Groundwater Levels. This method can be used to confirm flow-monitoring results for a particular area of a system by relating the groundwater table level to the elevation of sanitary sewers. If the groundwater table is below sanitary sewers, infiltration will most likely not exist within that area during certain times of the year because water tables can fluctuate above and below a sewer on a regular cycle.

Interviews. Local contractors who do a considerable amount of underground utility work in the area of interest are a knowledgeable source for groundwater information and should be consulted to determine areas of severe groundwater problems. Infiltration cannot exist without a continuous water source.

Municipal personnel can also be a valuable source of information on locating general areas of infiltration. Besides providing groundwater information, municipal personnel can provide information on infiltration problems such as backed up sewers and flooded basements.

Visual Inspection. Visual inspection can be done in different ways. One way would be opening manholes within a system during low-flow periods, typically at night, and visually judging depth of flow in the sewer. A more accurate means would be to enter the manhole and measure the depth of flow. Typically, on the upstream portions of a system, this method can be used to eliminate many areas by the existence or non-existence of significant flow during the early morning hours.

Visual inspection can also be accomplished by means of a television camera. Because of the cost involved, this generally becomes the final inspection procedure during a sewer system evaluation. Television inspection, however, is probably the best way to locate specific points of infiltration, particularly when performed during high groundwater periods. It is difficult to estimate quantities of infiltration by using a CCTV picture, and they are best measured by direct flow-measuring techniques.

COST EFFECTIVENESS. Cost effectiveness cannot be determined by simply measuring infiltration and computing incremental infiltration rates. U.S. EPA has attempted to define various incremental infiltration rates to indicate if removal of infiltration is cost effective, and these estimates can be used as a general rule of thumb. However, the only way to determine the cost effectiveness of infiltration removal is by conducting an analysis for each sewer system.

EVALUATION OF INFLOW

Indirect inflow is a delayed reaction to rainfall, primarily attributable to wet weather percolation through the soil to defective pipes, joints, connections, or frame-chimney joints or manhole walls. The nature and magnitude of inflow is heavily dependent on local conditions, including soil types, rainfall characteristics, and the depth and conditions of the existing sewer system.

Because the reaction to rainfall is so dependent on local conditions, comparing rainfall reactions from one system to another is virtually impossible. In fact, collection systems may react differently to similar rainfall events, depending on preceding weather conditions. As an example, the wastewater flow reaction to a specific rainfall would probably differ greatly if the rainfall was preceded by several weeks of dry weather as opposed to several weeks of wet weather. Therefore, a universal correlation of measured rainfall to inflow entering a particular system is almost impossible without many years of historical data.

Another important factor in inflow quantification is surcharging. Regardless of the amount or intensity of rainfall, once a system surcharges, the maximum peak inflow rate has been achieved for that particular storm. The peak inflow rate is ultimately controlled by the size and head of the leaks. Some systems surcharge rapidly and require only marginal rainfall events to reach maximum peak inflow rate, while other systems react slowly and surcharge only during prolonged, heavy rainfall. This is another indication that each system is unique and must be evaluated separately based on actual hydraulic conditions rather than data from systems in different locations.

To accurately quantify extraneous flows entering any collection system, total flows must be measured or accounted for and estimated, including contained flows remaining in the system and escaping flows such as overflowing manholes and bypasses. Complete and accurate flow monitoring is extremely important if I/I quantities are to be determined. Contained flows should be measured using one of the measuring techniques described in Chapter 6. Escaping flows also should be measured or estimated. If hydraulic conditions make measurement impossible, then the location, type, and duration of escaping flows should be noted and an estimate included with measured contained flows.

A common misconception in the quantification of inflow is that wastewater treatment flow records are the only data required for evaluation. In many facilities, flow measurement equipment is located on the effluent discharge, which would not record bypassed flows at the head of the facility, much less any overflowing manholes or other bypasses within the collection system.

APPLICABLE TECHNIQUES. Many types of flow measurement techniques, described in Chapter 6, are available and the selection of flow measurement equipment should be based on the type of information desired. For example, if flow data from various sections of the collection system are needed, then equipment that provides adequate data using manholes should be selected. Measurement can be made for entire collection systems or for a single leak, depending on the desired results.

A thorough investigation of overflows and bypasses should be discussed with appropriate employees or former employees to obtain information such as location, duration, type of overflow or bypass, and rainfall events necessary to produce such conditions.

If the location of bypasses (installed pipes to relieve surcharging or overflowing) are known before the contained flow measurement period, pipes can be plugged to allow for complete flow measuring. If plugging creates adverse conditions such as wastewater backups into residences, then the bypassing must be estimated or measured so that a bypass rate and quantity can be included in the final flow data.

Devices that adequately measure rainfall range from sophisticated electronic equipment to simple rain gauges. Rainfall data can be presented hourly, daily, or at whatever time interval is appropriate, depending on the equipment. Rainfall amount, duration, and intensity are important parameters that can be used independently or collectively to interpret inflow quantification.

Provided there are staff in the vicinity, the most economical rainfall measurement technique is reading a rain gauge hourly or daily. The gauge should be located in the area of the collection system under consideration for an accurate comparison of rainfall data to wastewater flows.

METHODOLOGY. If techniques can be defined as tools, then methodology can be defined as how those tools are used to compare, analyze, and evaluate collected data and other pertinent information to formulate results and conclusions. Wastewater flow data, measured rainfall, and existing conditions represent the tools. The type of wastewater flow data and rainfall data desired should depend on the financial resources available and to what extent the data will be used. A simple inexpensive study can relate daily rainfall to daily flows to determine if a problem exists and to what magnitude. More sophisticated and elaborate equipment will generate more extensive data, but also will cost more. Each community should implement a program to produce the minimum needed results to generate rehabilitation measures for the least cost.

Raw wastewater flow data can be converted to several useful flow parameters, including average dry weather flow (Q_D), dry weather peak flow (Q_{DP}), wet weather maximum daily flow (Q_{WM}), and wet weather peak flow (Q_{WP}). Other parameters can be developed on a case-by-case basis. The average dry weather flow is the average flow that occurs on a daily basis with no evident reaction to rainfall; dry weather peak flow is the highest measured hourly flow that occurs on a dry weather day; wet weather maximum daily flow is the maximum flow measured over a 24-hour period; and wet weather peak flow is the highest hourly flow measured during wet weather. Wet weather flows must include measured or estimated overflows and bypassed quantities.

Flow data also can be used to project annual flows if the study period is less than a year in duration. Flow data normally are presented through hydrographs that depict daily flows, hourly flows, and rainfall data as required. A typical hydrograph is shown in Figure 4.5. Note that the hydrograph includes escaping flows. The immediate return to dry weather flow range indicates inflow, whereas increased flows that remain elevated for several days usually indicate wet weather infiltration.

These basic flow parameters can also be used to quantify inflow. The difference in wet weather maximum daily flow and the average dry weather flow can be defined as the maximum daily inflow, and the difference in wet weather peak flow and the dry weather peak flow would be a conservative rainfall-induced peak rate.

Wet weather flow also can be compared to average dry weather flow to develop ratios that help define the magnitude of inflow problems during wet weather. Even in well-constructed, separate systems, the ratio of wet maximum to dry average (Q_{WM}/Q_D) usually ranges from 2 to 3, and the wet peak to dry average (Q_{WP}/Q_D) from 3 to 4; higher values indicate a more pronounced problem. These ratios only define magnitude and do not affect cost effectiveness. In some cases, ratios may be 2 to 3, respectively, and some rehabilitation may be required; in other systems, the ratios may be higher and rehabilitation may not be cost effective. The economics of rehabilitation are not based on flow, but solely on cost of transportation, treatment, and operation and maintenance.

Another important aspect of inflow is its correlation to rainfall events and durations. Because soil type, topography, area rainfall characteristics, and collection system conditions can vary significantly, no two collection systems operate identically, making it difficult to compare wet weather flow reactions of different systems. It also is difficult to predict a system's reaction to a particular rainfall event based on data from the reaction of another system to a similar rainfall event.

Even within the same system, identical rainfall events may produce different wastewater flow reactions. A certain rainfall episode could be preceded by several weeks of extended dry weather as opposed to several weeks of wet weather. In the first case, the ground is dry and capable of absorbing the rainfall into the soil matrix. In the second case, the soil probably is saturated and would have an entirely different effect on wastewater flows.

It is difficult to predict a flow reaction from a larger rainfall event based on data from a small event, as wastewater flows and rainfall intensity do not have a linear relationship. In most cases, there is a point at which all inflow

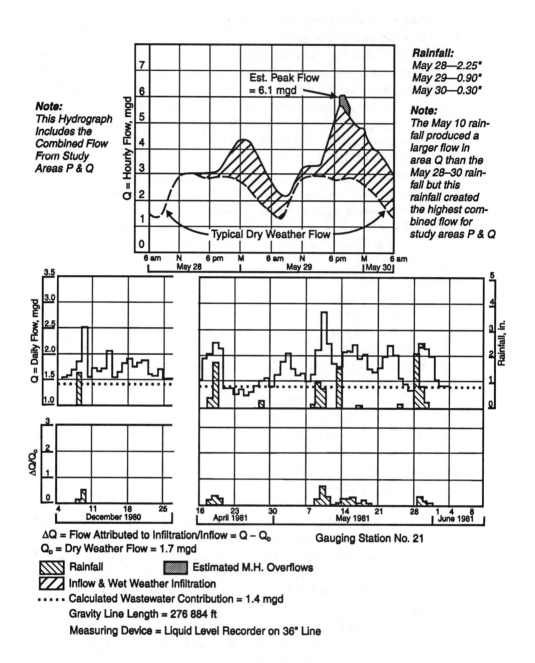

Figure 4.5 Hydrographs—the most common method of presenting flow data (in. × 25.40 = mm; mgd × 43.83 = L/s; ft × 0.304 8 = m).

sources are activated and a maximum inflow rate essentially is reached, regardless of the amount of rainfall. Extended rainfall can only increase the total quantity entering the system, which would be reflected in an increase in annual flows, but not the peak flow.

In summary, the quantification of inflow for entire collection systems or segments of collection systems can only be accomplished with measured flow data, including contained and escaping flows. Flow data must be correlated with measured rainfall and the evaluation process must be limited to the collection system in question.

RAINFALL-INDUCED INFLOW LOCALIZATION. Rainfall-induced inflow enters wastewater collection systems from three primary sources: collection lines, manholes, and service lines. Several different types of leaks can be associated with each primary source. These types of leaks include

- Collection lines
    - Broken or cracked pipe
    - Leak at a defective joint
    - Leak at a service tap
    - Multiple leaks along the entire line
    - Storm drain cross-connections

- Manholes
    - Through holes in the cover, between the lid and frame
    - Under the frame only
    - Within the top 0.6 m (2 ft) of the manhole chimney, corbel or cone
    - At the perimeter of the manhole bench
    - At a broken or missing manhole cover
    - At a broken or missing manhole frame
    - At a stub-out in manhole
    - Around pipes

- Service lines
    - Leak at a missing or defective cleanout or plug
    - Leak on a discontinued line
    - Storm drain cross connection
    - Foundation drains
    - Connected foundation drain
    - Roof drain connection
    - Surface drain connection

Other types of leaks may exist on the three primary sources, and the types of leaks will vary with local conditions. One part of the country may experience inflow primarily from drains connected to service lines, whereas other parts of the country may experience inflow from defective collection lines located in drainage ditches, rapidly draining soils, or defective frame seals. Inflow sources generally will be scattered throughout every collection system and cannot be isolated in small portions of the system with any degree of accuracy. Eliminating large portions of a collection system without smoke testing the area will reduce significantly the overall effectiveness of the sewer system evaluation.

Several techniques currently exist for locating rainfall-induced leaks, including smoke testing, dye-water testing, and leak quantification, as discussed earlier. Applicable techniques of inflow localization can be used independently or with one another.

COST EFFECTIVENESS. Regardless of the amount of extraneous stormwater that enters a sewer system or the quantity that can be identified for elimination (usually expressed as a percentage of the total I/I), the ultimate factor that defines excessive inflow is cost. Measured wastewater flows, including I/I, only establish survey and rehabilitation costs, construction and capital costs, and operation and maintenance costs. These cost parameters are used in the economical analysis to establish what percentage, if any, of the total inflow can be eliminated cost effectively.

Costs associated with survey and rehabilitation can be defined as survey tasks and type of rehabilitation. Inflow will also affect the need for lift stations, relief lines, and treatment units, and construction and capital cost of these items must be incorporated to the cost evaluation process. Operation and maintenance costs also will be affected by extraneous wet weather flows, pumping costs, energy usage at treatment facilities, and chemical usage.

A comparison of the cost to transport and treat rainfall-induced inflow with the cost to eliminate it allows the most cost-effective solution to be determined.

Figure 4.6 is a typical plot of a transportation and treatment cost curve, which includes operation and maintenance costs and the cost of construction of all facilities that are affected by inflow. This curve shows a reduction in overall transportation and treatment cost as the amount of inflow is reduced. Figure 4.6 also shows the survey and rehabilitation cost. As shown, the cost of repair increases significantly with the percentage of leaks eliminated. By combining these two curves, a composite cost curve can be developed. This curve has a minimum cost point that indicates the maximum amount of I/I that can be cost-effectively eliminated.

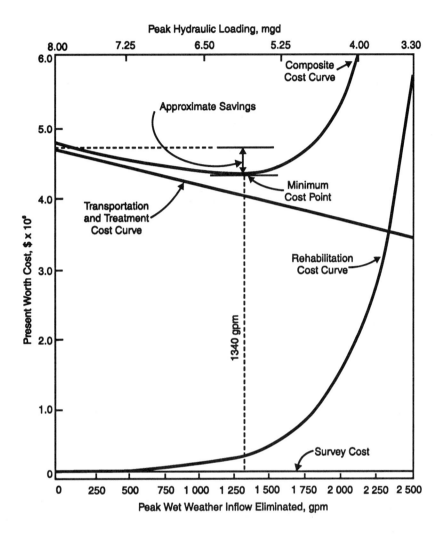

Figure 4.6 Cost curve for determination of inflow sources to be rehabilitated (gpm $\times$ 6.308 $\times$ 10^{-2} = L/s; mgd $\times$ 43.83 = L/s).

EVALUATION OF PHYSICAL CONDITIONS

STRUCTURAL INTEGRITY. When any evaluation is undertaken on a sewer collection system, an analysis for structural integrity and defects should be included. In many areas, the conditions of overburden may have become more than the original pipe or manhole wall was designed to accept because of changes in land use over the years, and some areas may have developed traffic loads unforeseen in the original sewer design. Pipe strength and manhole stability must be considered so the best alternative of rehabilitation can be selected. Depth limitations on pipes and manholes vary according to the type of material and its structural properties.

The second area of structural integrity involves the condition of the pipe itself, which can affect both hydraulic capacity and bearing strength of the pipeline. Each type of defect must be evaluated to determine its related impact on strength and capacity so the proper rehabilitation method can be chosen (see Chapter 3).

OPERATIONAL AND MAINTENANCE PROBLEMS. Every utility experiences some operational and maintenance problems attributable to specific causes. Problems and indicators of sewer pipe or manhole defects are as follows:

- Overflowing manholes;
- Sewer backup in buildings;
- Surcharged sewer;
- Exfiltration of wastewater;
- Sunken areas above the sewer line;
- Increased flows;
- Sand and gravel in wetwells;
- Stuck flap gates or open overflows; and
- Settlement of pavement around manholes.

Problems may be called in from the public or detected by routing inspection programs of the utility, and should be investigated. All information reported should be recorded by location so the problem history for a sewer line or manhole can be correlated with the specific manhole or sewer reach. Employees should be trained to collect the information needed for rehabilitation decisions. Most operation and maintenance employees concentrate on correction of the immediate problem without visualizing long-term rehabilitation solutions.

MAP CORRECTION AND UPDATING. Records indicating the skeleton of the piping and location of manholes are necessary for any good sewer collection system maintenance program. Each utility should have detailed plans of specific construction contracts so sewer lines and manholes can be located, and drawings should be to a specific scale so scaled measurement to manholes is possible from other structures in the area. In addition, there should be smaller scale drawings of the entire system so general patterns of the system are evident.

A specific procedure must be established by the utility for correction of errors and updating of drawings. Field personnel must be properly trained to recognize discrepancies between field conditions and map data and to record changes necessary to correct the existing mapping system. All corrections and changes should be coordinated by the centralized records section of the utility, and all corrections and changes should be accomplished in a timely manner with new drawings distributed to field personnel. The accuracy of drawings used by field personnel is critical to proper identification of sewer collection system components. Strict procedures for map formulation and updating may be cumbersome and time consuming for the utility but are critical for proper operation and maintenance of the sewer collection system.

The mapping system must properly identify sewer lines so detailed construction plans can be located quickly, and there should be a systematic method of identifying each manhole so sewer reaches can be identified individually. The identification of each manhole and sewer reach is imperative if the maintenance history and physical conditions are to be correlated properly. A method of identifying relocation work and rehabilitation work on sewer collection maps must be developed. By proper identification of all manholes and sewer reaches, a sewer system inventory can be developed and accurately maintained.

APPLICABLE TECHNIQUES. Visual Inspection. Visual inspection of manholes and pipelines can involve both a surface inspection program and an internal inspection program. Particular attention should be given to sunken areas over the sewer, areas with ponding water, condition at stream crossings, condition of manhole frame and cover, condition of any exposed brickwork, and whether manholes or special structures are visible.

The internal inspection program involves entering manholes and visually reviewing the condition of manholes, channels, and pipelines. Each manhole should be visually inspected for the following conditions:

- The condition of frame and cover;
- The seal of the frame to the manhole chimney;
- The condition of the manhole chimney and corbel or cone;
- The structural condition of walls;
- The joint condition in precast manholes; and
- Any visual leakage.

Each bench and channel portion of a manhole should be inspected for the following conditions:

- The condition of the manhole shaft;
- Any leakage in the channel;
- Any leakage between the manhole wall and the channel;
- Any damage or leakage where pipeline connects to the manhole; and
- Any flow obstructions.

Some pipelines can be inspected with lights or mirrors, though visual inspection in smaller sewers is limited in the scope of problems detected. Generally, the only portions of a sewer that can be seen in detail are those close to manholes. The remainder of the line can be inspected by mirrors to determine if there are any horizontal or vertical alignment problems or large leaks. Visual inspection of smaller sewers will not yield definitive information on cracks or other structural problems, though it does provide information needed to make rehabilitation decisions. Larger sewers can be visually inspected by walking through them.

Television Inspection. Closed-circuit television equipment has become the most effective method of inspecting the internal condition of sewers. Not only can reports be generated with the inspection, but a permanent visual record can be made for subsequent review. The range of inspection is almost unlimited, and small cameras will even televise 100-mm (4-in.) service laterals.

Operators using television equipment must be trained to obtain the required information. When recording data on report forms, operators must be able to translate the visual picture to a physical work description. Closed-circuit television operators must review conditions from the same perspective so results from different scans can be compared.

Both black and white and color CCTV units are available. Color yields additional depth perception, though black and white pictures give plenty of information and are used extensively for detection of physical conditions in sewer lines.

Television inspections should be recorded so they easily can be referenced to particular sewer reaches. In addition, if a sewer reach is videotaped, a cross file must be established for future review. Items that may be recorded on a television inspection report are as follows:

- Length of section;
- Type of pipe;
- Joint spacing;
- Root penetration (location and quantity rating);
- Grease sediment presence (location and estimated depth);
- Horizontal deviations (location and length);
- Open joints (location and severity);
- House connections (number, location, and condition);
- Water levels;
- Circumferential cracks (location and severity);
- Longitudinal cracks (location and severity);
- Missing pipe (length and severity); and
- Estimated I/I rate for all leaks or defects noted.

REFERENCES

Nogaj, R.J., and Hollenbeck, A.J. (1981) One Technique for Estimating Inflow with Surcharge Conditions. *J. Water Pollut. Control Fed.*, **53**, 491.

U.S. Department of Transportation (1988) *Manual on Uniform Traffic Control Devices for Streets and Highways*. U.S. Gov. Printing Office, Washington, D.C.

Water Environment Federation (1992) *Wastewater Collection Systems Management*. Manual of Practice No. 7, Alexandria, Va.

Water Pollution Control Federation (1969) *Design and Construction of Sanitary and Storm Sewers*. Manual of Practice No. 9, Washington, D.C.; Manual of Engineering Practice No. 37, Am. Soc. Civ. Eng., New York, N.Y.

Chapter 5
Combined Sewer Overflow Pollution Abatement

86 History and Problems
86 Regulatory Perspective
87 Planning
87 Issues
87   Environmental Setting
88   Regulatory Climate
88   Infrastructure
88   Sociopolitical Setting
89 Data Requirements
89 Analytical Tools
89 Investigation
90 Mapping
90 Field Investigations
90 Flow Monitoring
91 Sampling
91 Modeling
92 Quantity and Quality
92 Model Building and Verification

93 Assessing Performance
94 Determining the Pollutant Load
94 Identifying the Location and Causes of Deficiencies
94 Upgrading and Treatment Options
94 Source Control
95 Sewer System Control
95 Attenuating Peak Flows
95 Increasing System Capacity
96 Real-Time Control
96 Treatment Strategies
96   Screening
97   Sedimentation
97   Chemical Treatment
97   Filtration
97   Air Flotation
97   Disinfection
98 References

Abatement of combined sewer overflows (CSOs) presents the engineer or systems manager with one of the greatest rehabilitation challenges. Many variables increase the complexity of CSO control programs, and often the engineer or systems manager grapples with finding a cost-effective solution to the problem. Solutions can be interrelated with infiltration and inflow (I/I) and structural problems dealt with in this manual, and this chapter discusses some of the problems found in combined sewer systems. The subject is dealt with more extensively in *Combined Sewer Overflow Pollution Abatement* (WPCF, 1989) and in *Sewerage Rehabilitation Manual* (1986).

Approximately 1 100 sewer systems throughout the country are classified as combined systems and several others show significant adverse effects from increased flow during rainfall that are considered separate sanitary sewers. While this chapter is intended to address CSOs, it also pertains to separate sanitary sewers.

Rehabilitation of any sewer affected by local and upstream CSOs will vary depending on the respective pipeline hydraulic capacity. Hydraulic modeling analysis of the upstream tributary network should provide the information needed to determine pipeline limitations. It also is necessary to determine storm-influenced flows and ascertain surcharging or overflows and their annual frequencies. Determination of these events provides information on viable rehabilitation, replacement, or supplementing of existing systems.

HISTORY AND PROBLEMS

The first drainage systems were installed centuries ago to carry stormwater runoff. Wastes that accumulated around crowded urban dwellings subsequently were added to existing drainage systems in cities, and these combined sewer systems transferred street surface problems to nearby waterways. Receiving water bodies could not absorb the increased organic loadings and became heavily polluted.

Combined sewer systems then evolved to systems of interceptor sewers that conveyed dry weather flows to wastewater treatment plants. During peak wet weather storms, overflows went to receiving water bodies because collection systems were sized to carry only dry weather flows. The belief was that wet weather storm flows were diluted enough so they would not seriously affect the receiving environment. However, these overflows have been proven to have negative effects on the environment, and this problem has been a source of increased study.

REGULATORY PERSPECTIVE

The first federal legislation to address the problem of CSOs was the River and Harbor Act of 1899. This exempted from regulation "... refuse ... flowing from streets and sewers and passing therefrom in a liquid state." The Federal Water Pollution Control Act Amendments of 1972 constitute the current authority for national water pollution control programs. Federal funding to mitigate CSOs presently is not available except when a governor petitions the

U.S. Environmental Protection Agency (U.S. EPA) on the grounds that correcting the state's discharges is a priority.

Many nonpoint pollution sources are still beyond the technology of effective control, and sewer separation is viewed as an expensive mitigation measure. The current regulatory environment has not undertaken a full-scale attack on CSOs because of the enormous cost of abatement programs and the uncertainty of control measures. There is, however, a recognized need to address the problems of pollutants associated with CSOs, particularly because of their significant effect on bays and estuaries.

The intent of this chapter is to discuss reducing or eliminating CSO pollution from sewer systems. Cost-effective strategies exist in the form of structural solutions, nonstructural solutions, or a combination of the two.

PLANNING

Combined sewer overflows are randomly occurring events with variable pollutant loads, and their effects are difficult to assess. Planning for CSOs involves establishing plan objectives, considering available resources, and selecting the most appropriate analytical tools to formulate the best alternatives. Objectives must be clearly defined but often are difficult to set because of these variable conditions.

Because of the complex nature of CSO abatement, plan development often is undertaken in phases that confirm, broaden, or revise initial plan objectives. This section discusses plan objectives, data requirements, and selecting analytical tools. Plan objectives can be difficult to establish but are essential to a successful CSO abatement program. Data requirements provide a perspective for quantifying the translation of loading to impacts, and analytical tools directly affect plan development.

ISSUES. Relevant issues in CSO control such as the environmental setting, regulatory climate, infrastructure, and sociopolitical setting are identified when defining the abatement program and influence the plan objectives.

Environmental Setting. Receiving water is often the primary focus when defining the environmental setting. Important characteristics include loading size, beneficial use, type of water body, seasonal changes, and physical factors.

Characterizing the aquatic life focuses on the types of pollutants and the significance of impacts. The importance of seasonal impacts, transient variances in dissolved oxygen, type and diversity of other organisms required, and relative importance of the physiology of the water body should be assessed.

Knowledge of local meteorology is important because CSOs are a direct result of precipitation. Rainfall intensity causes specific CSO events based on the capacity of the sewer system. Statistical analyses of rainfall events can predict annual probabilities of occurrence.

Most collection systems have constraints that influence the extent of environmental improvements. Bottom deposits from historical overflows are often toxic and difficult to remove, and pollution from municipal and industrial wastewater treatment plants, stormwater discharges, and agricultural runoff also are significant constraints. Stream physiology, navigational uses, and recreational uses also affect CSO program approaches.

Regulatory Climate. Federal regulations on CSO control are imposed indirectly through the National Pollutant Discharge Elimination System program. The objectives of this regulatory program are to (1) ensure that CSO discharges occur only as a result of wet weather; (2) ensure that CSO discharge points will comply with the technology-based requirements of the Clean Water Act; and (3) reduce water quality, aquatic, and human health effects. State regulations vary across the U.S., although most states do not have specific standards for CSO discharge or treatment.

Infrastructure. Most combined sewer systems are between 50 and 100 years old, and structural integrity and conveyance capacity are important in developing a CSO abatement plan. Structural assessment will provide information needed not only to decide whether or not to separate the sewer system, but also to estimate the financial resources required to maintain the present level of service.

Increasing conveyance capacity could have a major effect on developing a plan for CSO abatement. Constructing relief sewers can be financially overwhelming and can alter the configuration of CSO abatement facilities. Integrating priorities to protect aquatic life, water body use, and property and to mitigate flooding is essential to defining the plan objectives.

Abatement facilities for CSOs affect existing wastewater treatment plants, and it is recommended to include the effects on existing treatment facilities in the overall CSO program. Construction activities will interfere with existing utilities, roadways, and flood control structures. Improvements from a CSO reduction plan will significantly affect existing infrastructure.

Sociopolitical Setting. Public awareness and political climate are important factors in defining plan objectives. The plan approach must respond to public input because of the public's role in financing the plan, and must respond to positions taken by elected officials. Responding to complex public questions is a key element in the overall plan. The uncertainty of the sociopolitical setting requires flexibility and recognition that the approach may need to be altered as the plan progresses.

DATA REQUIREMENTS. A review of the environmental, regulatory, infrastructure, and sociopolitical settings will define what must be addressed in the CSO abatement plan. Understanding these settings is essential to establishing priorities, evaluating alternatives, and determining the financial impact.

Data requirements will vary significantly depending on the approach selected for the CSO program. The extent and details of data needs and the analytical tools to be used are primarily determined by the selected CSO effects to be mitigated.

For some water bodies, the most significant effect of CSO pollution is from annual loads rather than a single CSO occurrence. Characterization of CSO impacts for critical events is important when threshold pollutant loading can be related to meeting a plan objective. Pollutant loads also can be estimated using a continuous loading characterization, which is normally on an hourly basis for long-term historical records of precipitation.

Effects of CSOs on receiving water bodies can be classified as either transient or long term. Transient effects are associated with a change in the concentration of constituents in the water column, and are important where water quality standards have been set or where short-term stress or aquatic life is important. Long-term effects are associated with accumulated settled solids or increasing concentrations of nutrients in the water column and are a major concern because of the pollutants associated with settleable solids. Extensive data are needed to predict a water body's response to transient and long-term effects, depending on the objectives of the particular CSO abatement plan.

ANALYTICAL TOOLS. Selecting analytical tools before establishing the plan objectives can limit the plan development process. The approach to plan development is then set by the tools rather than the objectives.

Planning often is an interactive process. Initial assumptions of planning objectives are based on a general understanding of the CSO effects, which leads to selecting certain analytical tools. Further study may shift the plan objectives, which affect the analytical tools selected.

Monitoring, sampling, and modeling are examples of effective analytical approaches. Specific discussions of these planning elements are presented later in this chapter.

INVESTIGATION

Understanding hydraulic and structural characteristics of the existing combined sewer system is a basic component of any CSO abatement plan and often requires the most time and money. Field investigations are expensive and often equal actual design costs. Not understanding the particular conveyance system could result in an incorrect remedial design approach.

MAPPING. Drainage area mapping is required when maps do not exist or are incomplete or outdated. All information about the combined sewer system should be compiled, as drawings and original construction records contain information that sheds light on the condition of the existing system. Reports describing the drainage area should be reviewed; aerial photographs provide information about land use and population densities; and interviews with current and former public works personnel can provide information for document data gaps.

The reliability of the information is critical, and a log of original information that has been verified through field investigation must accompany mapping documentation. New information must accurately be produced and incorporated to the verified original documentation.

FIELD INVESTIGATIONS. Field investigations provide the most accurate means of determining the operational status and condition of a combined sewer system. Available funds may dictate the extent of the field investigation because combined sewer systems may be too vast to inspect. Often, a priority system related to the plan objectives is needed and agreed to during the scope of work discussions. Field investigations can be performed by municipal staff or an outside contractor, depending on available equipment, staffing, budget, and time.

Field investigations should confirm the drainage area, identify locations for installing monitoring equipment, determine the structural integrity of the system, and assess the condition of mechanical facilities and the operational performance of the overall system.

FLOW MONITORING. Flow monitoring confirms the hydraulic characteristics of a combined sewer system and is used as a calibration tool for mathematical computer models. Flow monitoring may also be required for particular discharge permits. Some combined sewer systems are configured so that obtaining accurate flow data is difficult; accuracy, reliability, and cost separate good flow monitoring from bad.

Flow measurement techniques used in wastewater collection systems also are used in evaluating combined sewer systems. Flow-monitoring systems range from simple depth-measuring equipment to sophisticated combined velocity and depth-measuring equipment. Simple depth-measuring equipment measures only depth, using floats, pressure transducers, bubblers, or ultrasonics.

More sophisticated, velocity sensors are used to measure either a point velocity or velocity across a section of flow and they help resolve surcharge and backflow conditions that invalidate depth sensors. Many new flow monitors measure both depth and velocity in a single piece of field equipment.

Most recent CSO instrumentation was based on depth and discharge data. Communities and agencies are now adding rainfall data to the list of controls.

Rainfall duration and intensity are valuable to the CSO evaluation model. Rainfall data coupled with flow-monitoring results establish the characteristics of the CSO system.

SAMPLING. Information on the quality and quantity of overflow is needed to characterize the existing combined sewer system. Pollution abatement of CSOs primarily is aimed at preventing the degradation of receiving water from pollutant discharges during rainfall events. It is vital to know the constituents of the overflows and their relative pollutant loadings. Combined sewer overflows normally consist of sanitary wastewater, settled waste and slime scoured during high flows, and rainfall, all of which flow over the ground surface before entering the sewer.

Abatement planning for CSOs must both manage the flows and ensure the removal of pollutant loadings as required to meet water quality standards. Appropriate sampling and analysis provide data for producing mass balances and inputs to water quality models that are the basis for design criteria for CSO abatement facilities.

MODELING

With the development of several comprehensive computer models for analyzing the quantity and quality of the flow in combined and storm sewer systems, modeling has become an integral part of most CSO abatement studies. Modeling not only allows the simulation of existing conditions but also provides a means to test alternatives and controls before implementation in the field. However, the models are only a sophisticated representation of the physical reality, and many caveats and cautions apply.

Modeling has been useful in analyzing particular problems in combined sewer systems, including CSO effects on the receiving water quality, flooding and hydraulic problems, I/I problems, and in assessing the effects of other rehabilitation efforts, including structural relining.

The overall objective of a modeling study often is clearly defined, but the investigation of one problem may lead to another set of issues and problems. Three levels of modeling can be identified: planning, designing, and operational control. Less detail can be used in planning the models to provide a cost-effective comparison of general abatement strategies and risk assessment. More detailed models can be used for analyzing existing drainage systems or proposed pollution control alternatives. Modeling also can be used to achieve real-time operational control to manipulate a sewer system, thereby avoiding harmful overflows to receiving waters.

QUANTITY AND QUALITY. Many models exist to convert rainfall to runoff and to perform flow routing, leading to a reasonably accurate prediction of a runoff hydrograph that gives volumes and peak flows. The most significant input parameters are catchment imperviousness and the rainfall hyetograph, with parameters such as slope or roughness being less sensitive. However, any simulation is only as good as the data used and the accuracy of the model compared to the actual system. Model calibration and verification using measured flows are important if the results are to be used confidently.

Quality modeling is not possible without calibration and verification data, which require costly monitoring programs. Quality modeling, therefore, is used only when absolutely necessary and when requisite calibration and verification data are available. Accurate quantity modeling is often used to determine the CSO volumes and hydrograph peaks. This information may be enough to suggest CSO abatement strategies, although a demonstration of receiving water quality benefits is becoming more common before funding for options is authorized. Modeling water quality parameters with the sewer system is often a prelude to receiving water quality modeling. Statistical, data-based approaches to uncertain model predictions also have been used.

MODEL BUILDING AND VERIFICATION. The hydraulic model of the system has four main roles: (1) to help understand the circumstances that lead to recorded hydraulic problems; (2) to help establish target performance criteria for pollution; (3) to allow system performance to be compared to the required performance criteria; and (4) to provide a basis for assessing the effect of rehabilitation.

The model should meet several criteria. It should be able to simulate the performance of the existing system and appurtenances, simulate surcharge, predict flooding, be flexible in use, and not involve excessive cost related to the scale of the system and its problems.

The hydraulic investigation is concerned primarily with the performance of the core parts of the combined sewer system, and the model should represent these parts of the system in detail. The remainder of the system needs to be represented only in enough detail to ensure accurate simulation, to which end some large catchments may be subdivided. Peripheral areas modeled in a simplified form should have uniform topographic and hydraulic characteristics, no significant ancillaries, and a simple layout.

Before any model is used to design major upgrading measures on an existing combined sewer system, the simulated performance should be checked to ensure it represents the actual performance of the system. This process is called calibration or verification. Actual performance is assessed from two broad types of data: (1) historic data from surcharging or flooding records, preferably related to rainfall data; and (2) field measurements, ranging from relatively simple observations of surcharge depth in manholes to sophisticated measuring and logging equipment that record either depth only or depth and velocity of flow. The use of rainfall measurement devices is an essential part of a field measurement program.

After obtaining system performance records, model simulations are compared to recorded events. If suitable agreement is not obtained, model input data should be checked, followed by sewer records. It may be necessary to confirm possible causes of error by a site inspection.

ASSESSING PERFORMANCE. The objectives of assessing the quantitative performance of a system are to establish the return frequency of the design storm corresponding to the onset of surcharge and flooding and to determine performance characteristics of ancillary structures during a historical series of rainfall events. The model is run for a series of standard-profile synthetic design storms to establish the return frequency and duration of storm critical to the system. The accuracy of the modeling results will decrease as the severity of rainfall events increases because the model may be run under conditions more stringent than those used for calibration and verification. A number of factors may cause this, including the operation of unknown features, a small error becoming progressively more significant, errors in the modeling of the overland routing hydrology, inaccurate modeling of complex surcharged conditions, or inaccuracies induced by the necessary simplification of the model.

When investigating quality problems, the model can be run for a time series of rainfall events, which will define acceptable discharge levels in terms of polluting loads and the frequency or duration of spills. Recent U.S. EPA requirements specify a minimum spill frequency, such as once per year, although the duration and acceptable quality parameters are likely to be catchment and site specific.

There are fundamental differences in how a model is used to assess quantity and quality. When assessing quantity, a series of design storms of varying return periods are run through the model to determine when and if individual pipe lengths begin to surcharge or flood. Assessing quality is more involved, with ecologists defining the nature of existing problems and determining targets for upgrading. In most cases, the model must be run for a historic series of rainfall events that represent the types of storms that cause frequent CSOs. This realistic running indicates the frequency and volumes passing to receiving waters, information essential to meeting U.S. EPA guidelines for spill frequency.

Time series rainfall events are a sequence of rainfall events that statistically represent the long-term rainfall patterns for a given geographic area. Such rainfall input is necessary when investigating quality problems because CSOs often occur more frequently and during smaller rainfall events than the design storms that cause surcharge and flooding.

DETERMINING THE POLLUTANT LOAD. Ideally, a deterministic model should be used to relate spill volume to pollutant load in the discharge from a CSO. However, when such a model is not available, sampling can establish the true average value for the quality of the base flow, with the quality of stormwater being calculated by applying a series of multiplying factors. The phenomenon known as "first flush" should be considered when applying these factors, particularly in flatter catchments where sewers may not have self-cleansing velocities. When sampling is not viable, a second approach using tabulated data from other sources may be used.

IDENTIFYING THE LOCATION AND CAUSES OF DEFICIENCIES. After assessing the performance of a combined sewer system, it is necessary to identify the causes of deficiencies. By comparing flows and capacities within individual pipeline lengths and from system results as a whole, it is possible to establish whether problems on individual or combinations of pipeline lengths are caused by the inadequate capacity of the lengths or by effects in other locations. The volumes discharged through CSOs should have been determined through modeling and can be converted to pollutant loads, which are then considered in light of receiving water quality objectives and discharge frequency requirements.

*U*PGRADING AND TREATMENT OPTIONS

Hydraulic modeling allows the engineer to assess and develop options for addressing problems identified in the drainage area. These options are designed to satisfy the economy of construction, reduced disruption and environmental effect during and after construction, and the division of the work into stages. As much of the existing combined sewer system as is practical should be retained, and maximum use should be made of its capabilities. Many of the options described below rely on one or both of these criteria.

SOURCE CONTROL. Source control involves reducing intentional and unintentional hydraulic inputs to the sewer system. Options include disconnecting separate sewer areas from the combined sewer system, diverting peripheral areas to other catchments where additional capacity is available, providing alternative drainage for large impermeable areas, disconnecting roof leaders from the combined sewer system, and reducing I/I. The principal advantage of this approach is that it addresses the problem at the source and deals with prevention rather than a cure. However, it is unlikely to overcome major deficiencies without major expense.

SEWER SYSTEM CONTROL. Increasing the use of the existing sewer system involves increasing its capacity by removing throttles, cleansing the sewers, and reconstructing or relocating CSOs. The removal of local throttles deals with local problems cost effectively by making full use of downstream capacity, though the operation may result in transferring problems downstream.

Cleansing removes silt and debris, which reduces the effective capacity of the system and may resolve local problems. However, cleansing may be costly and detrimental to a sewer in poor condition, and may become a routine maintenance problem.

Reconstructing CSOs provides structures with good hydraulic control and can reduce pollution or spill volume to receiving waters. It may be difficult, however, to achieve acceptable water quality objectives and spill frequency requirements.

Relocating CSOs relieves the sewer system at the most appropriate point, but the preferred location may not be acceptable from a practical, environmental, or political standpoint.

ATTENUATING PEAK FLOWS. Flow attenuation can be achieved by providing water storage inside or outside the sewer system or by using excess storage capacity with strategically placed throttles or regulator stations.

In-system water storage reduces the required capacity downstream, although it may be difficult to find suitable sites for potentially large tanks. Off-system storage also reduces the required capacity downstream and is inexpensive to apply in new developments, although it may be difficult to apply in existing areas because of maintenance complications. Throttles, in the form of orifices, penstocks, pipes, and regulator stations, effectively control flow and mobilize the available in-system storage capacity. Flow attenuation may involve surcharging the upstream pipes in the system and is limited to sewers in which the risk of surcharge is acceptable. Ongoing inspection and maintenance costs may be high.

INCREASING SYSTEM CAPACITY. Probably the most frequently used and traditional option for CSO abatement programs is to increase the sewer system capacity either by replacing existing sewers with larger capacity sewers or by reinforcing the system with additional sewers.

Open-cut replacement removes problems on the pipeline length and is the traditional method. However, it is disruptive in downtown areas and expensive, particularly when deep. Tunnel replacement also removes the problems on a pipeline length, in addition to reducing disruption and providing flexibility on the line and level. However, it is usually more expensive than open-cut methods and may require high-level collector sewers.

Reinforcement using relief sewers and interceptors relieves hydraulically overloaded sewers without requiring construction. The sewers are often tunneled, reducing disruption, but the method is expensive and may require additional overflow, treatment, or storage capacity at the wastewater treatment facility.

The advancement of trenchless construction methods, such as horizontal directional drilling and microtunneling, and trenchless replacement methods, such as pipe bursting, may improve the acceptability and cost effectiveness of replacement and reinforcement options for smaller diameter sewers, up to approximately 1 m (3 ft) in diameter. Details are found in Chapter 7.

REAL-TIME CONTROL. Real-time system control for reducing CSOs is a recent development and is becoming a more attractive option with the advent of more powerful computerized monitoring and control systems. The objective is to adjust gate positions and pump speeds during a storm event to increase the use of the storage in response to changes in rainfall patterns and flow conditions within sewers.

Real-time control systems monitor and control the regulator in response to rainfall events in the entire collection system, not just the area of the regulator. Control systems also can monitor rainfall at various locations, measure or calculate flows, and adjust setpoints according to programmed strategies and are more complex and sophisticated than individually controlled facilities and more involved than static control.

In most real-time control systems, control decisions are made by computers rather than people. In a computer-controlled system, a preprogrammed computer decides whether and how regulators are to be operated, though it is supervised by operators and can be taken over manually at any time.

TREATMENT STRATEGIES. Treatment can be used with or without other types of CSO abatement, depending on the water quality problems to be solved. Treatment options can be located wherever effluent leaves the collection system and enters a receiving body of water. It is possible, for example, to treat an existing or new CSO to improve the quality of the discharge. If an existing wastewater facility is being expanded, it may be economical to use existing treatment technology.

Several types of treatment technology are viable options for a CSO abatement program: screening, sedimentation, chemical treatment, filtration, air flotation, and disinfection.

Screening. Screens separate solids from the wastewater stream and are designed to remove a given particle size, although solids smaller than the screen size can be removed as the screen becomes clogged. However, screens have relatively high maintenance costs and are susceptible to clogging, tearing, and mechanical failure.

Sedimentation. The gravitational removal of suspended solids from CSOs can be used with CSO facilities. Sedimentation structures also can provide additional storage capacity and can function at various flow rates. Chemicals can be added to provide higher suspended solids and biochemical oxygen demand (BOD) removal, and the facility can be used as a point to provide disinfection, reducing coliform discharges.

Treating effluent at an existing wastewater treatment facility can be one of the most cost-effective solutions for reducing CSO effects. Existing facilities should be examined for acceptable operating efficiencies and upgraded where appropriate. Secondary treatment processes, including contact stabilization, trickling filters, rotating biological contactors, and treatment lagoons, can be used to remove nonsettleable colloidal and dissolved organic matter. However, operational problems can arise when treating intermittent storm events by biological processes. Storage of flows often is necessary to control influent and maintain a uniform flow rate to the treatment system. Biological treatment of CSOs is considered viable only if treatment is possible for both wet and dry weather flows.

Chemical Treatment. Coagulants to enhance sedimentation can be used with other forms of CSO control, such as screening, filtration, and dissolved air flotation.

Filtration. Periodic backwashing of the filter medium effectively removes from 50 to 80% of suspended solids without adding polymer for CSO applications. Up to 60% of BOD can be removed by adding polymer.

Air Flotation. In this method, air is introduced to the liquid stream, forming bubbles and allowing the air and discrete solid aggregate to be removed for CSO treatment. Operating this type of facility may be more difficult than the operation required for sedimentation. However, flotation can accept higher hydraulic loadings than sedimentation, and small and light particles that settle slowly can be removed quickly and more completely.

Disinfection. Disinfection reduces the concentration of pathogens in wastewater before it is discharged to receiving waters. Disinfection can remove almost 100% of the total coliforms in CSOs and should be used with some form of upstream solids removal to be cost effective. The long contact times involved in conventional wastewater treatment are not appropriate for treating CSOs because of the high flow rates and dilution involved. Treatment can be achieved by providing an increased disinfection dosage with intense mixing to ensure disinfectant contact with the maximum number of microorganisms.

Disinfection reduces potential public health hazards from CSOs, but used alone does not improve water quality or aesthetic impacts. Dechlorination facilities may be required where chlorine is used as a disinfectant.

*R*EFERENCES

Sewerage Rehabilitation Manual (1986) (Addenda published in 1990). 2nd Ed., Water Res. Cent., Eng., U.K.

Water Pollution Control Federation (1989) *Combined Sewer Overflow Pollution Abatement*. Manual of Practice No. FD-17, Alexandria, Va.

Chapter 6
Flow Monitoring

100	Purpose	110	Velocity
100	General Data Needs	111	Manual Measurements
100	Maps of Study Area	111	Weirs
101	Records	111	Triangular (V-Notch) Weir
101	Sewer Maps	112	Rectangular (Contracted) Weir
101	Sewer Conditions		With End Contractions
101	Flow Records	112	Rectangular (Suppressed) Weir
101	Bypass and Overflow Information		Without End Contractions
101	Emergency Pumping	113	Trapezoidal (Cipolletti) Weir
102	Water Usage	113	Compound Weir
102	Rainfall Data	114	Flumes
102	Groundwater, Lake, and Stream Data	115	Parshall Flume
102	Demography	116	Palmer–Bowlus Flume
102	Recordkeeping Program	117	H-Flume
103	Monitoring Program	117	Trapezoidal Flume
103	Research	117	Dye, Chemical, and Radioactive
104	Site Selection		Tracers
104	Key Manholes	118	Pumping Station Calibration
105	Bypasses	118	Calibration Using Wet Well
105	Overflows		Drawdown or Return
105	Factors Affecting Flow	119	Calibration Using a Velocity Meter
106	Equipment Selection and Sizing	119	Calibration Using Discharge
106	Timing and Data Correlation		Volume
107	Safety Program	120	Calibration Using a Pump Curve
107	Monitoring Duration	120	Recording Calibration
108	Instantaneous Monitoring	121	Bucket Test
108	Random Monitoring Interval	121	Stage Measurement
108	Continuous Short-Term Monitoring	121	Automatic Flow Meters
109	Continuous Long-Term or Permanent	121	Depth Recorders
	Monitoring	121	Probe Recorders
109	Available Equipment	122	Bubbler Recorders
109	Measurement Theory	122	Pressure Sensors
110	Manning's Equation	122	Float Recorders
110	Calibrated Discharge Curves (Stage	122	Ultrasonic Recorders
	Discharge Curves)	122	Capacitance or Electronic Recorders

123	Velocity Meters	129	Measuring Rainfall
123	Doppler Meters	129	Data Analysis
123	Orifice, Nozzle, and Venturi Meters	129	Editing Data
123	Current Meters	129	Filtering
124	Velocity Probes	129	Offset
124	Site Work	129	Time Adjustment
124	Maintaining Metering Equipment	129	Viewing Data
124	Manual Measurements	129	Determining Key Values
126	Continuously Recording Measurements	130	Quality Assurance
		130	Manual Verification of Meter Data
126	Depth-Velocity Meters	130	Review of Recorded Data
126	Depth-Only Meters	130	Problem Areas
128	Measuring Groundwater	131	References

PURPOSE

Flow monitoring establishes the basis for any sanitary sewer rehabilitation program. Monitoring work should be conducted first and in a manner that will provide reliable, accurate data. Failure to obtain good data will hinder or prevent the success of a rehabilitation program.

Flow in a sanitary sewer system consists of base flow, infiltration, and inflow, which are described in Chapter 4. Flow monitoring provides the data necessary to quantify these components.

The process begins by dividing the collection system into basins, mini-basins, or areas. The number and location of meters and meter sites should be chosen to make the analysis of data as easy as possible. For example, flows to be metered should be measured only once. The flow-monitoring process will provide flow volumes by location. This allows for pinpointing problem segments of the study area.

Flow monitoring is conducted during wet weather periods to measure collection system responses to weather-related sources of infiltration and inflow (I/I). When conducting a monitoring program, it also is necessary to monitor and correlate rainfall to collection system flows.

GENERAL DATA NEEDS

MAPS OF STUDY AREA. It is essential to have maps of an area before starting any system study. Useful information may be obtained from U.S. Geological Survey (USGS) topographic maps, state organizations such as natural resources departments, geologic surveys, health departments, regional planning organizations, county governments, sewer districts, and utility companies.

A review of these data before or early in the study program serves as a valuable orientation to the area and helps identify potential trouble areas, such as altered drainage paths, filled lakes and streams, and areas of high groundwater.

RECORDS. Organizations with responsibility for the sewer system have some type of recordkeeping system, ranging from handwritten notes on scrap paper to computerized data retrieval systems. Regardless of how the records are kept, it is essential that they be collected, sorted, reviewed, and used for a meaningful and effective program.

Sewer Maps. A complete, up-to-date set of sewer maps is essential to a proper sewer system study. Maps should be to a scale convenient for both office and field use.

Sewer Conditions. Information about the sewer condition includes such items as date and methods of construction, type of pipe, manhole, and joint material, sewer areas, and chemical damage areas. This information should be supplemented by type and frequency of complaints from the residents and meetings with maintenance personnel, who may be able to define problems and answer questions because of their working knowledge of the system.

Flow Records. Sewer flow information should be collected, verified, and analyzed. Even if manual flow readings at the treatment plant of less than daily frequency are the only flow information collected, a wealth of information may be found in years of continuous flow data.

Occasionally, subsystem flows may be determined from pumping station records and instantaneous or short-duration flow records from prior metering programs. Large industrial or commercial users within the study area may have sewer discharge records available.

Bypass and Overflow Information. Most sewer systems have some type of bypass or overflow. Bypasses and overflows usually are constructed to relieve a flooding or backup condition, though they can act as a natural system relief where raw wastewater is discharged from a manhole cover because of severe system surcharging.

Both bypasses and overflows must be documented as to location and frequency of discharge. During some portion of the system study, quantification of their discharge is needed to accurately assess the sewer flow conditions.

Emergency Pumping. In many sewer systems, emergency pumps may operate during storms or high-water conditions to prevent backup and overflows. Quantification of the pumping activity gives the analyst a better understanding of the functioning of the collection system. During a flow-monitoring program, the time, duration, and quantity of emergency pumping must be determined and considered to establish true system flows.

Occasionally, construction within the study area can have associated pumping or dewatering activities that affect sewer flows by reducing groundwater levels. In many instances during such activities, discharge is pumped directly to a sewer manhole, legally or illegally, and can drastically distort flow in that area.

Water Usage. The quantity of water used by residential, commercial, and industrial users usually is documented by water meter readings. Usage records are helpful in determining expected wastewater flow in the collection system (base flow). Although not always available and often distorted by private water supplies such as wells and stream intakes, base flow information is needed to understand the collection system.

Rainfall Data. Rainfall data for the area covered by the collection system can be useful in determining relationships between sewer flow events. Rainfall also can help explain unusual flow situations present in any system study. Amounts of rainfall and the variations across the study area are an essential element of any flow-monitoring program.

The type of rainfall records available may vary from an official weather station to a rain gauge at the treatment plant, and it often is necessary to establish a rain gauge network as part of the sewer system study.

Groundwater, Lake, and Stream Data. Groundwater, lake, and stream elevations can play important roles in I/I quantities, and any record of these elevations is important in understanding sewer system operation. These records may be collected by a variety of organizations, including the USGS, U.S. Army Corps of Engineers, state agencies, city and county governments, and industrial and contracting firms. The news media and private individuals also can help, particularly concerning record high-water levels.

Demography. Demography and population trends within the collection system study area also must be considered when analyzing records. Past and projected population and population trends may help explain historical flow data and changing flow patterns. These data also should be considered when developing the study plan so that adequate data are gathered to allow proper planning for future population expansion and shifts.

Recordkeeping Program. Even under the best of circumstances, the existing records and recordkeeping program may not provide all the needed information. Consequently, there is an opportunity to reassess the current data collection and storage program, and often only minor revisions to the existing program are warranted. However, a consistent recordkeeping program may need to be initiated. Even if a specific study is not in progress, it is essential that a routine, meaningful recordkeeping program be maintained.

MONITORING PROGRAM

The complexity of a system-monitoring program should only reflect how much information is needed to accomplish the job. Before determining the scope, the team designing a program should review the following questions:

- How many data are required?
- How accurate must the data be to satisfy the goals?
- Can the required data be found in existing files?
- What type of system is to be monitored?
- How experienced and knowledgeable are the personnel?
- What equipment is available?
- In what condition is the equipment?
- How much flow monitoring has been conducted previously on the same system?
- Are data from previous work applicable to the proposed program?
- What is the project duration?

These questions should be asked frequently before and during the monitoring effort to keep program goals clearly defined.

RESEARCH. The foundation of a good flow-monitoring program is the research conducted before designing the program. The knowledge of maintenance staff should be used as fully as possible, as experiences of people working on the system for several years through varying climates and changing conditions often can be used to identify specific locations where flow conditions are causing problems. Frequently, the most expensive short-term monitoring program does not locate such items.

Pump records in the system should be reviewed. Often, such records are only in terms of hours of pumping as documented manually by maintenance personnel. In such cases, the pumps must be calibrated to determine their true capacity. Ideally, all pumps in the system are less than 10 years old and have well-calibrated flow meters and recorders on the discharge force main. This type of situation provides a consistent, long-term database if recorder charts are saved and filed properly.

Another valuable source of flow-related data is the wastewater treatment plant. In addition to reviewing flow records at the plant, the researcher also should secure the influent biochemical oxygen demand (BOD) and suspended solids concentrations to determine the dilution of the flow at various times. Treatment plant personnel should be interviewed regarding flow patterns and general history and so that a rapport for coordination during the monitoring program can be established, as specific rates of pumping may be necessary during monitoring.

One frequently neglected database is the long-term history of water and wastewater flows in the area. Simply plotting the average monthly water and wastewater flow values parallel with each other on the same time scale for a few years can reveal much about historical flows in the system. For example, a long-term increase in water production may indicate an increase in population or a change in water consumption habits; it can also mean the water system is getting old and leakage is becoming a major problem. Also, a steady increase in wastewater with no accompanying increase in water production can indicate either a deteriorating sewer system subject to more I/I or a few years with above average rainfall.

When other long-term plots are added, such as precipitation, temperature, and BOD, this data source can be informative and should be analyzed early in the program.

When maps and other physical information about the system are secured and reviewed with the maintenance staff, copies should be available for indicating problem areas experienced by staff. Other data sources for the monitoring program are as follows:

- Soil borings from projects in the area indicating groundwater levels;
- Geological information suggesting conditions such as perched water tables;
- Older topographic maps before the sewer system was established, indicating natural drainage patterns, old swamps, or old stream beds filled in during development;
- Storm sewer maps showing the density of the drainage system and areas with inadequate storm drains (sanitary sewers in such areas are particularly subject to I/I); and
- Design drawings, showing trunk sewer profiles pieced together.

SITE SELECTION. When the researcher has collected existing information on the sewer system, these data should be catalogued and reviewed. If existing flow data are insufficient, then a monitoring program should be designed.

The type of system being monitored dictates the type of equipment and procedures used. A large, urban combined sewer system requires a different approach than a small suburban subdivision. Thus, site selection should begin with the size of the pipes and the system in mind.

Key Manholes. The sewer system should be divided into subsystems, each generally consisting of approximately 6 100 m (20 000 ft).

Exact sizes of subsystems are determined when key manholes are located. Manholes where the long-term (control) monitoring will take place should be on straight sections of pipe with consistent slope upstream and downstream, cleaned with the sewer, large enough for easy access, relatively dry, free or relatively free from surcharging, located in highways or streets free of heavy traffic, and otherwise accessible. Finding such manholes can often prove difficult, necessitating compromise on the size of subsystems.

If information is available, it often is helpful to divide subsystems according to the way the sewers were constructed. A pattern with all sewers in one subdivision isolated in one subsystem can be used to find whether sewers in one subdivision are more susceptible to I/I than those in another. Otherwise, the selection of key manholes is left to the judgment of program planners.

When key manholes and major subsystems are located, it may be necessary to subdivide each of these subsystems into smaller areas to provide more useful data. An economic evaluation of performing other investigations and the magnitude of the problems in an area will determine the length of these smaller subsystems.

Bypasses. Bypasses not identified at the beginning of the program can cause many data interpretation problems after monitoring is completed. If the area takes in more than one sewer district, upstream districts may be bypassing flow to a stream or another sewer system. Other bypassing is typical in an older system where newer trunk sewers were installed to intercept flow from certain points in the older system. If these points are not documented, they can be overlooked in a later monitoring program. For best results, the program should include monitoring bypasses to establish total flow within the system.

Overflows. A combined system may have numerous combined sewer overflows. Also, in older sanitary sewer areas with insufficient drainage, overflows may have been installed to relieve flooding. If inflow is a monitoring program concern, significant overflows must be monitored.

Overflows may direct flow to waterways, other subsystems, or storm sewers. They are sometimes difficult to access, and frequently it is difficult to monitor flow through an overflow unless there is a manhole or pumping station on the overflow pipe.

Overflows need to be monitored to establish the total system flow if meaningful flow data are to be obtained. Otherwise, the whole program could be rendered ineffective because of incomplete data.

Factors Affecting Flow. When observing flow in a sewer, the researcher should never assume theoretical conditions. Field conditions offer variables which must be uncovered and taken into account. A manhole at a change of slope can cause distorted depth and velocity readings, especially if the change causes turbulence at the manhole.

Each type of monitoring equipment will have distinct advantages and disadvantages under various field conditions. Equipment installation and monitoring results may be affected by some common conditions. For example, deposition in the invert of a sewer can cause distorted depth of flow data and result in unnecessary turbulence and flow meter malfunction. Deposition should be removed before equipment installation.

A break in the service connection in the section downstream from a monitor can cause slight surcharging in the key manhole, distorting flow data. Slipped joints can lead to deposition and surcharging similar to breaks in service connections. Dips in the sewer caused by settling or poor construction can also distort flow data. Broken pipe, depending on the severity of the break, can cause a number of problems. Minor cracks can allow significant infiltration, and major collapses can cause flow stoppage.

The effects of such conditions on the flow-monitoring program can be reduced if the sewers upstream and downstream of the proposed key manholes are investigated. If any of these conditions are severe, another manhole should be considered.

Many other conditions can cause monitoring problems. Water consumption is critical and should be analyzed to approximate wastewater flow. The local climate, season, and resultant precipitation or freezing can upset the program.

Tidal conditions can upset flow monitoring if overflows to a tidal basin are being monitored. Depending on the situation, tides can cause reverse flows in the overflow pipes. Wind-induced currents on large bodies of water, such as lakes, can have a similar effect. These reverse flows distort the flow picture.

Equipment Selection and Sizing. Monitoring program equipment is expensive and should be investigated thoroughly before purchase, lease, or use. Many meters require batteries, bubblers require a supply of compressed air or nitrogen, and tanks must be rented and refilled periodically. Certain spare parts, such as pens and probes, should be purchased with the unit, and extra meters should be available to serve as exchange meters if a particular monitor becomes damaged or malfunctions. Otherwise, key data may not be obtained simply because of improper program planning. Other considerations are difficulty of calibration, susceptibility to moisture (submerging), and sturdiness.

Most types of monitors are adaptable to large or small sewers, though manufacturers' specifications and recommendations should be reviewed, considering the particular metering location, conditions, and requirements.

Timing and Data Correlation. Two of the most critical factors in managing a flow-monitoring program are timing and data correlation. The only realistic method of ensuring that all flow-monitoring data can be correlated for the same intensity, duration, and frequency of storm is for all monitoring sites to be measured simultaneously under the same storm conditions. However, this may not be feasible because of the size of the system or other considerations, and may still present correlation problems because of geographical variations in rainfall and rainfall patterns.

The most common method of trying to handle data correlation for different rainfall events is to provide the system with a series of control monitors, which should be installed on a permanent basis for at least the duration of the total flow-monitoring program in the collection system. It is best to install them permanently and let them remain in the system before beginning a study. This will allow their use during the flow-monitoring period for data control and correlation of the effects of different storm events on the system. It also will allow control of work to be accomplished during other phases of a study program and for continuing system maintenance after any special, short-term study is completed. However, they must be properly maintained and calibrated if their data are to be meaningful. Permanent monitors are useful in scheduling rehabilitation activities, checking the effectiveness of rehabilitation, scheduling maintenance, determining I/I trends, and providing data concerning future area development.

SAFETY PROGRAM. Safety too often is played down or ignored as a nonproductive aspect of the flow-monitoring program, but there are at least three reasons why it should command a high priority:

- Proper safety can prevent loss of personnel time, thus saving money and preventing program disruption;
- A strong safety program will convince field personnel of the manager's concern for their welfare and will improve morale; and
- An inadequate safety program can be an invitation to a liability suit following an injury.

An adequate safety program is essential. It is not sufficient to merely provide personnel with equipment; they must be trained to use the equipment and be constantly reminded of the need for safety (WPCF, 1983).

MONITORING DURATION. The duration of the monitoring period at a particular manhole is defined by the data required. If dry weather flow pattern assessment is the only need, 1 week of monitoring may suffice. If it is desirable to monitor particular flow conditions in the sewer or to determine what flow is caused by a certain type of storm event, several weeks of monitoring may be necessary. Even after that period, the budget for the program may be used up before ideal conditions are realized. The following comments about monitoring should prove useful to the program manager.

Instantaneous Monitoring. Instantaneous monitoring often can be used to gain valuable information. In some cases, instantaneous monitoring may be all that is available because of budget or time constraints.

It is useful to spot check for depth and velocity in the sewer at the start of the program to check dry weather flow and get an idea of flow extremes. In combination with cup gauges, instantaneous depth and velocity checks can be used to define the range of flow in the sewer. The correct type of monitoring equipment then can be selected to match flow conditions. Spot checks should be used during the program to recalibrate continuous monitoring equipment.

An additional use of instantaneous depth and velocity measurements is to monitor a single storm event. If sufficient personnel are available and there is sufficient prior warning of the storm event, teams can be stationed at several locations to measure the event. Spot depth and velocity measurements then can be taken at prearranged times (every 15 or 30 minutes) during the storm event. To ensure success, personnel must be well trained and available for the event, each team at a manhole must be equipped in advance, and equipment used by one team must be calibrated with equipment used by all other teams.

Random Monitoring Interval. Random monitoring intervals need only be of the length required to monitor desired flow conditions. A random interval program frequently is the result of a tight equipment budget. Then, because equipment is not available to monitor flow in all key manholes over the same period of time, individual devices must be moved such that equivalent conditions are monitored at all sites. The 1-in-6-months storm may occur during the first week of monitoring at points 1, 2, 3, and 4. After moving equipment to points 5, 6, 7, and 8, however, an equivalent storm may not occur for months. Rain gauges and permanent monitors should be installed so that some means of correlation may be determined. Additional engineering time for correlation may offset some of the reduced cost of monitoring for this type of program, and if it is necessary to monitor the 1-in-6-months storm flow at all locations, the budget should be flexible. If not, some compromise may be necessary regarding the desired data.

Continuous Short-Term Monitoring. It frequently is desirable to monitor flow at each key manhole continuously for a short period to gather specific data for a study. Again, if the equipment budget is sufficient and the timing is right, all locations need simultaneous monitoring for only a short time. In general, approximately 14 to 20 days should be considered a minimum duration under optimum precipitation conditions. The actual monitoring time must be determined based on weather and rainfall conditions and data needs, and sometimes a monitoring duration of several weeks or months is required.

Continuous Long-Term or Permanent Monitoring. Occasionally it is necessary to establish semipermanent or permanent monitoring at a specific site, as when a special district is discharging to another district. Also, it may be desirable to monitor the effectiveness of flow reduction during a large-scale rehabilitation program; determine collection system capacity data, I/I trends, and scheduled system maintenance; or investigate system expansion. In such cases, equipment costs may be significant and it may be necessary to modify an existing manhole or even build a new chamber especially suited for the new equipment.

Evaluation of equipment before purchase cannot be overemphasized, particularly for a permanent monitoring site. The literature should be reviewed, other municipalities consulted, vendors brought in, and lists of installations secured. Several installations should be contacted to check the reliability of the equipment, and nearby installations should be visited to compare conditions with those at the proposed site.

*A*VAILABLE EQUIPMENT

Monitoring flows within a wastewater collection system is important. The resulting data form the basis for determining user costs, volume of rainwater or groundwater entering the system, existing line capacity, treatment plant operations and maintenance, effectiveness of the rehabilitation program, and design of future needs.

Many techniques exist to measure sanitary sewer flows. The equipment and technique selected should depend on the resources available, the degree of precision required, and the physical conditions within the sewer. This section presents a brief summary describing the techniques of the more commonly used flow-monitoring methods.

MEASUREMENT THEORY. The two principal methods for measuring flowing fluids are the direct-discharge method and the velocity-area method. For the direct-discharge method, the rate of discharge relates to one or two easily measured variables. The velocity-area method requires knowledge of the cross-sectional area of the conduit to be metered and the velocity of flow. Sanitary sewers generally operate only partially full and are classified as open channels. As described in Chapter 4, measurement of flow in the simplest terms takes the form of

$$Q = AV \quad (6.1)$$

Where

Q = flow, L/s (cfs);
A = cross-sectional area, m^2 (sq ft); and

V = mean velocity, m/s (ft/sec).

Manning's Equation. The familiar Manning equation for open channel flow is as follows:

$$V = (k/n) R^{0.67} S^{0.5} \tag{6.2}$$

$$Q = (k/n) R^{0.67} S^{0.5} A \tag{6.3}$$

Where

- n = coefficient of channel roughness, dimensionless;
- R = hydraulic radius (area divided by wetted perimeter), m (ft);
- S = slope of energy gradient (S = slope of line only for uniform flow), m/m (ft/ft);
- A = cross-sectional area, m^2 (sq ft); and
- k = 1.00 for metric units, 1.486 for English units.

Three values must be determined before a flow rate can be computed: depth of flow, pipe roughness, and energy gradient (slope) where depth is measured. Not only must these values be determined, but their variability with relative depth also must be determined (Gutierrez and Siu, 1982). It is a common misconception that Manning's n is a constant; actually, the roughness coefficient varies with liquid depth.

Calibrated Discharge Curves (Stage Discharge Curves). It is possible to use a depth recorder and obtain fairly accurate data by calibrating the flow-monitoring location. This method relies on the depth recorder for the source of raw data and for the calibration of flows at various times and liquid levels using instantaneous weir measurements or other primary calibration devices. Depth is noted and a calibrated flow rate observed at the same time. By making observations at various liquid depths, a discharge curve can be developed to convert depth readings to calibrated flow rates. This method is not valid if the pipe surcharges or is subject to varied backup conditions from downstream obstructions or high flows entering the line downstream.

Velocity. The direct measurement of velocity in a wastewater system can provide accurate and reliable flow data. A velocity reading and depth of flow can be used to calculate flow rates by using the continuity equation.

Average velocity can be determined by either the one- or two-point method. In the one-point method, the velocity meter is placed at the 0.6 depth and the measurement is used as the average velocity. The one-point method should be considered a rough approximation and used only where time or shallow depth of flow are significant factors. The two-point method records velocity at the 0.2 and 0.8 depths, and the average of these two readings is taken to represent the average velocity.

MANUAL MEASUREMENTS. Manual methods are the most widely used techniques for measuring instantaneous or short-term flow. Generally, equipment is portable and flows can be determined immediately using published curves, nomographs, or tables.

Weirs. The weir is a common device for measuring wastewater flow because of its ease of installation and low cost. To design a weir that will provide useful data, the following should be considered:

- The weir should be constructed of a thin plate with a straight edge or a thick plate with a knife edge;
- The height of the weir from the bottom of the channel to the crest should be at least two times the expected head above the crest (tends to lower approach velocity);
- All connections between the weir plate and channel must be watertight;
- To prevent a vacuum on the nappe underside, the weir should be ventilated;
- The weir must be level;
- The weir crest and approach channel must be cleaned periodically;
- The head over the weir should be measured at a point located upstream of the weir at least 2.5 times the head over the weir at the weir plate or in a stilling well as long as the point of measurement is not affected by the drawdown of the water level approaching the weir crest;
- Weirs should be placed in a straight stretch of sewer with little slope to lower the approach velocity (it may be necessary to use baffles for uniform velocity distribution upstream);
- The weir size and type should be determined only after a preliminary field survey has determined existing and anticipated flow rates;
- Weirs installed under field conditions should be calibrated to ensure accurate measurements; and
- Under surcharge conditions or when free fall over the weir does not occur, weir equations and data should not be used.

Measurements are taken by recording the head above the weir crest and determining flow rates by calculation, nomographs, or tables. Figure 6.1 presents commonly used weirs. The recommended usage for each weir is described below.

TRIANGULAR (V-NOTCH) WEIR. This type is good for accurate measurement of low flows. It is the best weir profile for discharges of less than 30 L/s

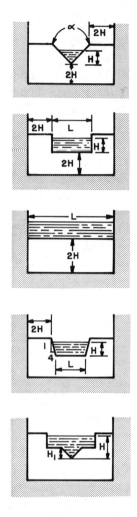

Figure 6.1 Flow-monitoring weirs (from top: triangular, rectangular contracted, rectangular suppressed, trapezoidal, and compound).

(1 cu ft/sec) and may be used for flows up to 300 L/s (10 cu ft/sec). Its general operating range when installed in a manhole is 0 to 90 L/s (3.0 cu ft/sec).

RECTANGULAR (CONTRACTED) WEIR WITH END CONTRACTIONS.
This type is able to measure higher flows than the V-notch weir and has a more complicated discharge equation than other weirs. It is widely used for measuring high flow rates in channels suited to weirs.

RECTANGULAR (SUPPRESSED) WEIR WITHOUT END CONTRACTIONS.
This type is able to measure the same range of flows as a contracted rectangular weir but is easier to construct and has a simpler discharge equation.

However, use is restricted as the width of the crest must correspond to the width of the channel. It may be difficult to obtain adequate aeration of the nappe.

TRAPEZOIDAL (CIPOLLETTI) WEIR. This type is similar to a rectangular contracted weir except that the inclined ends result in a simplified discharge equation.

COMPOUND WEIR. This type is a combination of any two types or sizes of the above weirs to provide a wide range of flows. It has an ambiguous discharge curve in the transition zone between two weirs.

Weirs can be used with depth recorders to record head (H) over the weir, though care must be taken to ensure the head being recorded is carefully referenced to the weir crest. Also, caution must be used when monitoring flows in sewers. Solids and debris tend to settle upstream of the weir, which may lead to odor and corrosion, may affect the weir length, and may alter the accuracy of the measurements. Figure 6.2 shows an example of rectangular weir construction and installation. Commercially constructed weirs are readily available for most standard sites and locations. Advantages and disadvantages of weirs are listed below.

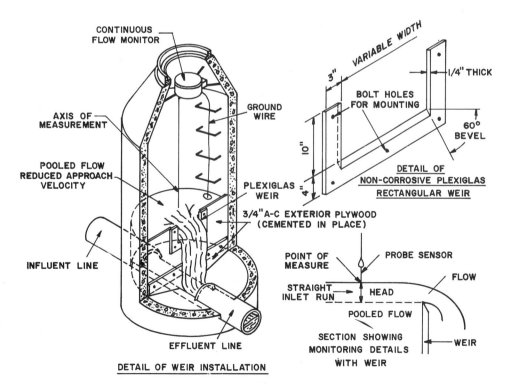

Figure 6.2 Weir details and installation (in. × 25.40 = mm).

Flow Monitoring

- Advantages
  - Low cost
  - Easy to install
  - Easy to obtain flow by standard equations and nomographs

- Disadvantages
  - Fairly high head loss
  - Must be periodically cleaned; not suitable for channels carrying excessive solids
  - Accuracy is affected by excessive approach velocities and debris
  - May be difficult to make accurate manual measurement in sewers because of limited access

Flumes. Flumes operate as open channel forms of the Venturi tube. In a flume, constriction at the throat forces flow to go through critical depth, followed by a hydraulic jump if the slope allows subcritical (low-velocity) flow. There are several types of open channel flumes, including the Parshall, Palmer–Bowlus, H-Flume, and trapezoidal configurations. Flume hydraulics are discussed in Metcalf and Eddy, Inc. (1981). Flumes generally are capable of providing results accurate to within 3 to 5% (Kulin and Compton, 1975, and ASCE, 1976). Some advantages and disadvantages are as follows:

- Advantages
  - Self cleaning to a certain degree
  - Relatively low head loss
  - Accuracy is less affected by approach velocity than in weirs
  - Data are easily converted to flow using tables or nomographs

- Disadvantages
  - High cost
  - Difficult to install

Flow moving freely through a flume is calculated from a measurement of upstream water level. Depth-measuring devices can be used to obtain continuous flow data; manual measurements can give instantaneous flow data. Flumes are self cleaning with no sharp edges to cause settling or clogging, and approach velocity affects their accuracy less than other measuring devices. These monitoring devices are used widely in measuring influent and effluent flows at wastewater treatment plants.

A flume should be located in a straight section of the open channel, without bends immediately upstream. The approaching flow should be well distributed across the channel and relatively free of turbulence and waves.

Generally, a site with a high approach velocity should not be selected for a flume installation. However, if the water surface just upstream is smooth with no surface boils, waves, or high-velocity current concentration, accuracy may not be greatly affected by the velocity of approach.

Consideration should be given to the height of the upstream channel and its ability to sustain the increased depth caused by flume installation.

Although less head is lost through flumes than over weirs, it should be noted that significant losses may occur with large installations.

The possibility of submergence of the flume caused by backwater from downstream should be considered, although the effect of submergence on the accuracy of most flumes is less than with weirs.

PARSHALL FLUME. The Parshall flume (Figure 6.3) was originally developed for monitoring irrigation flows. It can be formed from many different

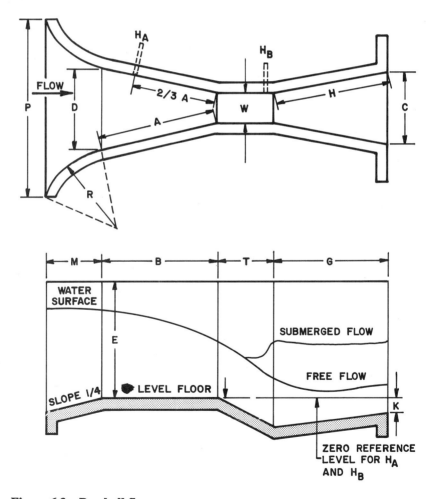

Figure 6.3 Parshall flume.

Flow Monitoring

materials and is commercially available in metal and reinforced fiber glass with throat width ranging from 25 mm to 15 m (1 in. to 50 ft).

This type of flume commonly is installed in wastewater treatment plants for flow monitoring and has performed well in these locations because of accurate results and low head requirements. Because the channel is rectangular and a head drop of at least 70 mm (3 in.) is required for free flow, it is difficult to install a Parshall flume in existing sewers for I/I monitoring.

PALMER–BOWLUS FLUME. The Palmer–Bowlus flume (Figure 6.4) is effective in measuring sewer system flows. An open-channel flume, it is easily adaptable to a wide variety of sewer diameters, though a separate unit is required for each sewer diameter. The basic hydraulic principle is the same as that for a Parshall flume. Constriction in the side and a step up in the sewer bottom causes the flow to go through critical depth, which can then be related to the flow rate. Development of a rating curve is shown by Metcalf and Eddy, Inc. (1981).

The Palmer–Bowlus flume is available commercially in several materials and configurations. Different types can be used for direct installation in sewers or for I/I studies in manholes and can be temporary or permanent.

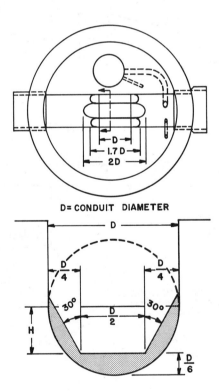

Figure 6.4 Palmer–Bowlus flume details.

The main advantage of the Palmer–Bowlus flume is that it is easily adaptable to the circular section of sewer lines. It also introduces a minimal head loss, which prevents surcharging of sewers. Major disadvantages are that it may not be as accurate as the Parshall flume and usually has a low capacity.

H-FLUME. The H-flume is useful for measuring flows over a wide range. Parshall and Palmer–Bowlus flumes have useful ranges of flow variations of about 10 to 1, but the H-flume can be used to measure flow variations of 100 to 1. Most convenient for measuring stormwater or combined sewer flows, the H-flume requires a free discharge at the outlet, thus a substantial head loss is introduced.

TRAPEZOIDAL FLUME. A trapezoidal flume is suited for monitoring small flows and has often been installed in earthen or concrete ditches. The flume channel has a flat bottom that reduces the opportunity for silt accumulation, and head loss is small.

Dye, Chemical, and Radioactive Tracers. The dye-dilution technique is a simple, potentially accurate, and quick method for the determination of flows in sanitary sewers. Flows can be measured even under partially full or surcharged conditions without entering manholes. This method normally is used to obtain instantaneous flow rates, but with added equipment can be used to monitor flows on a continuous basis. Advantages and disadvantages include the following:

- Advantages
  - No entering of manholes
  - Saves time and provides instantaneous flow data on many sewer sections
  - Independent of sewer site, dimensions, velocities, and surcharging
- Disadvantages
  - Samples must be analyzed as soon as possible (most dyes decay in sunlight)
  - Temperature corrections may be required
  - Instrumentation is expensive
  - Dye is expensive
  - Needs at least 100 sewer diameters for dye mixing before sampling

Chemical and radioactive tracers can be used to measure wastewater flows. The methodology used in chemical and radioactive tracers is similar to dye-dilution except the initial concentration of the wastewater stream must be determined using Equation 6.4:

$$Q_s = Q_t (C_t - C)/(C - C_s) \tag{6.4}$$

Where

Q_s = stream discharge, m³/s (cfs);
Q_t = tracer discharge, m³/s (cfs);
C_t = initial concentration of tracer discharged, mg/L;
C = initial concentration of tracer in waste stream, mg/L; and
C_s = tracer concentration of sample, mg/L.

Dye, chemical, and radioactive tracers also can be used to measure velocity between two control locations. The time it takes for the center of gravity of an injected tracer to travel to a downstream control point is noted, and the velocity is computed by dividing the distance between the control points by the travel time. Average depth of flow then is used to calculate flow. Dye, chemical, and radioactive tracers, when used with care, can provide flow data accurate to 5% or better (Kulin and Compton, 1975, and ASCE, 1976).

PUMPING STATION CALIBRATION. Because of their location and the difficulty in monitoring upstream of them, pumping stations are logical sites for monitoring wastewater flows not measured by other methods. It is important to calibrate each pump and combination of pumps because the design capacity in most cases is not representative of actual field conditions. In a pumping station with two constant-speed pumps that alternate in operation and run together under high-flow conditions, three calibrations will be required. This technique is applicable when the force main discharges freely and not to another common force main shared by other pumping stations. Possible pump combinations become excessive and difficult to monitor when other pumping stations feed into a common, shared force main.

Calibration Using Wet Well Drawdown or Return. This method requires accurate dimensions of the wet well from as-built drawings or field measurements with locations of incoming sanitary sewer lines. The following procedure is used to calibrate each pump and combinations of pumps:

- Knowledge is needed of the pump controls, the depth to which the wet well can be lowered without exposing the suction lines, and the elevations of sewer lines entering the wet well.
- With all the pumps turned off and the wet well level below all incoming lines, measure the level of the liquid from some reference point.
- Start the pump and let it run for 1 minute (less time if the wet well empties too quickly), being careful to record the time accurately.
- Record the drawdown distance (D_d) from the reference point at the time (T_d) the pump is turned off.
- With pump(s) off, record the time (T_r) it takes to refill the well to a known distance (D_r).

- Refill the wet well and repeat the procedure until three consecutive measurements are the same.
- If there is more than one pump, repeat the procedure for each pump and possible combination of pumps. If a station has three pumps, calibration must be made on pump 1; pump 2; pump 3; pumps 1 and 2; pumps 2 and 3; pumps 1 and 3; and pumps 1, 2, and 3.

The pump flow rate can be calculated from Equation 6.5:

$$Q_p = [(A_w)(D_d)]/T_d K = [(A_w)(D_r)]/T_r K \qquad (6.5)$$

Where

Q_p = calibrated flow rate for pump, m³/s (gpm);
A_w = area of the wet well, m² (sq ft);
D_d = depth of drawdown in time T_d, m (ft);
T_d = time of draw down, s;
D_r = depth of return in time T_r, m (ft);
T_r = time of return, s; and
K = conversion factor, 1 for metric units, 448.8 for cu ft/sec to gpm.

Inconsistent results likely would be caused by fluctuations in the incoming flow or infiltration directly to the wet well. Should this occur, it may be necessary to undertake calibrations at night during low flows, temporarily plug the incoming line, or calibrate during low groundwater conditions. If the incoming lines are plugged, the refill time and distance measurements are omitted, so no return flow will be measured and the effect of incoming flow is zero.

Calibration Using a Velocity Meter. Velocity meters can be used to obtain the velocity in the force main while each pump and combination is calibrated. Sensors are placed on the discharge force main and velocity measurements are taken. Velocity meters should be located so bends and valves do not cause inaccurate measurements. Conversions to flow rates are simply calculated using $Q = AV$ (Equation 6.1).

Calibration Using Discharge Volume. For small pumping stations, the following procedure may be considered:

- Turn pump(s) off and plug the incoming line at the manhole downstream from the force main discharge point;
- Turn pump(s) on and allow the line to surcharge to depth D_o above the outgoing pipe (care must be exercised not to cause backup to basements or homes) and turn the pump(s) off;
- Turn on pumps to be calibrated for approximately 1 minute, then turn off and record the run time (T_r) and depth (D_1);

- Remove the plug and allow the surcharge level to decrease to a new D_o level;
- Calculate the flow rate

$$Q_p = (A[D_o - D_1])/T_r \qquad (6.6)$$

Where

Q_p	=	flow rate of pump, m³/s (cu ft/sec);
A	=	cross-sectional area of manhole, m² (sq ft);
D_o	=	initial liquid level before test, m (ft);
D_1	=	final liquid level at end of run time, T_r, m (ft); and
T_r	=	running time of pump(s), s.

- Repeat to obtain consistent measurements.

Calibration Using a Pump Curve. Pump curve calibration involves plotting the system curve, describing the head loss through the discharge piping with respect to flow against the pump curve supplied by the manufacturer. The intersection of the two points gives the flow rate at which the pumping system operates, which should be along a small section of the curve as the flow is reduced when the wastewater level in the well drops. The system curve should be plotted for both high and low wet well levels (pump on and pump off).

The accuracy of this method depends on obtaining reliable values for head loss through the system. Friction factors should take into account age, corrosion, and slime growths in the discharge and force main piping. Pump curves may also change slightly as the pump experiences wear. In general, this is not as accurate as some of the other calibration techniques.

Recording Calibration. After pumps and combinations of pumps have been calibrated, on-and-off cycles (events) of each pump should be recorded. Many pumping stations are equipped with meters on each pump that record the running time, which is used for maintenance but can be recorded hourly or daily to obtain flows (running time multiplied by the calibrated pumping rate is equal to the volume of wastewater pumped).

An inexpensive method of recording running time is to install electric clocks on the motor circuits so they operate only when the pump is on. The running time of the pumps can be recorded hourly or daily and converted to flows, and a clock should be installed that will operate only when both pumps are running to account for all possibilities.

Meters designed for pumping stations are available and greatly reduce installation and monitoring time. One event recorder can record on the same strip chart the time on and off of three or more pumps simultaneously. Other solid-state digital storage devices can store the same information in memory for easy retrieval without having to analyze strip charts. Special event recorders can record cycles on variable-speed pumps.

BUCKET TEST. The bucket test is probably the most widely used method of obtaining accurate, instantaneous flow rates for free-falling flow. This technique is limited to small flows where the discharge can be contained in a calibrated vessel. A bucket or container of known volume is placed to contain the entire flow, and a stopwatch is used to measure the time it takes to fill the container. The flow rate is calculated by dividing the measured liquid volume gathered by the elapsed time of filling. Several repetitions should be made to obtain consistent measurements.

STAGE MEASUREMENT. This device is relatively inexpensive and yields valuable information regarding extremes of flow depth. Cups are mounted on a vertical shaft at measured intervals and used to approximate high water levels in a storm. Such cup gauges are useful during the early stages of a study program to define flows in the system and determine the types of equipment required. They are also used to supplement other equipment during the formal monitoring program, particularly if the budget is tight. Such gauges must be checked and emptied after each storm or high-flow period.

AUTOMATIC FLOW METERS. Automatic flow meters continuously record various flow parameters with a minimum of labor. Data collected may be displayed, recorded on charts, stored on magnetic tapes or solid-state memory, or transmitted from the field to the office by telephone or radio.

These meters can save considerable time and effort compared to manually recording flow data, but proper installation, calibration, and maintenance require individuals with a basic knowledge of hydraulics and proper maintenance procedures for the meter in use. There are many automatic flow meters manufactured with various options and techniques for recording and analyzing flows.

Depth Recorders. Depth recorders are versatile and can be used to measure liquid levels in a pipe, head over a weir, depth in a flume, or other applications where unattended depth measurements are necessary. Common techniques of recording liquid levels (depth) are probe, bubbler, pressure sensor, float, ultrasonic, and capacitance or electronic.

PROBE RECORDERS. A probe recorder measures or senses the wastewater surface with a thin electrical wire. When the wire (probe) makes contact with the wastewater, a circuit is completed, and the probe stops and retracts slightly. Every few seconds, the probe senses the liquid level. Data are transferred to the recorder either aboveground or in the manhole, depending on the equipment installation. It is necessary to clean and calibrate the recorder during installation and periodically during the monitoring period by setting the recording device to the actual liquid level, measured by a ruler or other device. After initial calibration, the probe and recorder will sense the liquid surface level and record the depth on charts or other permanent devices.

BUBBLER RECORDERS. Whereas a probe recorder senses a liquid surface level, a bubbler recorder senses actual depth. A small pneumatic compressor, air tank, or canister supplies a low gas flow to the bubbler tube. The bubbler system then measures the water pressure at a selected point that corresponds directly to the water depth, and a sensitive pressure transducer makes the comparison between the force necessary to expel a bubble from the sensor tip and the atmospheric pressure. This pressure differential is then converted to a depth measurement and is recorded. Because the bubbler can be clogged with grease and other solids, it is necessary to clean the tubes regularly.

PRESSURE SENSORS. Pressure-sensing devices, which may be mounted directly under water, measure the fluid pressure directly on the sensing element and then electronically convert it to an equivalent depth of fluid. Field cleaning and calibration checks should be standard procedure.

FLOAT RECORDERS. The float recorder is a mechanical device, which, like the probe, measures the liquid surface level. A float with a mechanical pivot or cable and pulley converts the vertical rise and fall of the liquid surface to a continuous depth recording. In some instances, a stilling well is fabricated because of turbulence in the flow. The float method requires accurate calibration of the initial depth at the time of installation with periodic depth checks during the monitoring period and possible cleaning. Data can be recorded on charts as depth measurements or directly converted to flow rates, depending on the instrumentation.

ULTRASONIC RECORDERS. In this method, an ultrasonic device transmits a high-frequency pulse and the liquid surface level is obtained by measuring the difference in time or frequency of the reflected return signal. A comparison is electronically made and converted to a depth or, in the case of Doppler equipment, velocity measurement. Measurement of the air-to-liquid interface requires an initial calibration of depth, and data are recorded on a continuous basis in the form of depth or flow rates.

As with all automatic depth recorders, ultrasonic devices can be used effectively with weirs, flumes, or other primary measuring devices. Although a unit is not in contact with liquid or fouled by debris, discrepancies in the results can be caused by waves, foam, and other floating materials.

CAPACITANCE OR ELECTRONIC RECORDERS. Capacitance or electronic recorders use flowing liquid as an electrolyte to sense the depth of flow. The probe can be inserted to a flume or sewer invert, though because accumulation of grease and debris can cause false readings, field calibration checks and cleaning are recommended.

Velocity Meters. Automatic flow monitors that use velocity measurements can provide accurate data even under highly fluctuating liquid levels. Velocity may be automatically recorded with ultrasonic Doppler meters, magnetic meters, mechanical current meters, or other methods.

In most cases, a depth of flow is recorded with the velocity to use the flow equation $Q = AV$ where the area component is a function of the liquid depth. The advantage of obtaining a velocity component is that, even under surcharge conditions, accurate flow data can be obtained because the cross-sectional area becomes the area of the pipe being monitored. It is not uncommon for flows to reverse direction under surcharge conditions, and some velocity monitors will record a positive or negative velocity.

Weirs, flumes, and Manning's equation rely on open channel flow, and data should not be used under surcharge conditions. These data gaps can be eliminated by using a velocity recorder, manually obtaining velocities during the surcharge condition, or installing meter pairs in consecutive manholes to determine the energy gradient.

DOPPLER METERS. Velocity meters of the Doppler type usually are connected to the outside of the pipe to be monitored. This generally requires pipes that flow full (such as force mains) and have sufficient suspended solids for the transmitted signal to reflect back to a receiver. Absence of sufficient solids may be overcome by a bubble injection system, though this generally is not required when measuring wastewater. Ultrasonic meters can record flows in closed pipes without obstruction to the flow.

ORIFICE, NOZZLE, AND VENTURI METERS. These three types of flow meters are used for measuring flow in completely filled pipes. The basic concept is to form a constriction in the flow so velocity increases and pressure decreases.

There are many configurations for such meters, which can be installed in the pipe or at the discharge end. The main disadvantage to this type of meter is that the constriction allows solids to accumulate, particularly in an orifice meter. In addition, grease can interfere with proper operation.

CURRENT METERS. Portable, propellor-type current meters often are used to obtain velocities. Data are obtained from current meters by counting revolutions and then converting to velocity, though some meters convert and directly display the measured velocity. Care should be taken to ensure that suspended matter in the wastewater does not clog the meter and that there is sufficient depth of flow to obtain accurate measurements.

VELOCITY PROBES. Portable magnetic and Doppler ultrasonic velocity probes use no moving parts and are not affected by clogging. Inserted to the flow, magnetic probes consist of wire coils that generate an electromagnetic field and electrodes that measure the induced voltage created when a conductor (wastewater) passes through the field. The induced voltage is converted to velocity.

Doppler ultrasonic meters transmit a continuous high-frequency pulse to the wastewater. A reflected pulse of different frequency is received, and the difference between the two frequencies is proportional to the velocity. This in turn is displayed as a velocity measurement on the meter. Most ultrasonic probes are mounted on the outside of pipes, making them ideal for obtaining velocities in force mains, sludge lines, and other pressure pipes. These devices can produce accuracy of 1 to 5% (ASCE, 1976).

A hot-wire anemometer can also be used for spot velocity checks. The passing fluid causes a heated wire or probe to be cooled. The temperature change is proportional to the velocity and can be recorded electronically. These devices usually are delicate and more suited for laboratory work than measurement of sewer flow.

*S*ITE WORK

MAINTAINING METERING EQUIPMENT. Monitoring sites that are continuously recording during a flow-metering project should be checked periodically. A check of the monitoring site should be made the day after installation, and semiweekly visits are recommended to service the measuring device and collect data. A site visit after a major storm event is advisable to confirm meter conditions. Less frequent maintenance could diminish accuracy because of debris, damage to monitoring equipment, or equipment failure in the unpredictable environment of the sanitary sewer.

Manual Measurements. Manual measurements can be made with temporary or fixed devices, as described previously in this chapter. These result in immediate readings, which can be converted to flow rates. Figure 6.5 illustrates manual weir installation.

Manual measurements can help supplement data collected by continuously recording meters, and often are performed at night or during low-flow periods. Gauging performed while plugging upstream flow is called isolation and measurement (I & M) and is used to monitor one or two line segments. When four or more sections are monitored, plugging is not always used and measurements are taken at outlets to an interceptor or at junction manholes. These are called outlet and junction (O & J) measurements.

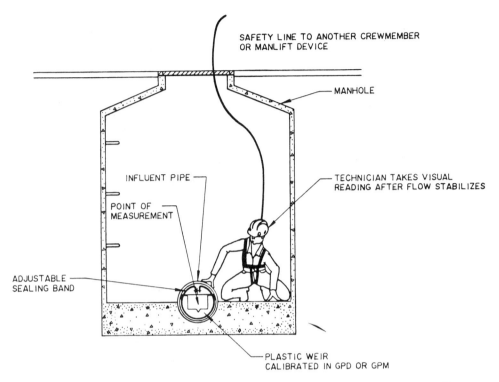

Figure 6.5 Manual weir installation (gpd $\times 4.381 \times 10^{-5}$ = L/s; gpm $\times 6.308 \times 10^{-2}$ = L/s).

When conducting I & M readings, upstream line segments are measured first. Before installing a weir in the downstream manhole, the plug or plugs should be installed at the upstream manhole. Plugs should be secure to prevent leakage or loosening from pressure buildup on upstream lines. The following are necessary elements of performing I & M:

- Plugging
  — Plugs should be secure and not leak.
  — The plug should be tied securely to the steps in the manhole to prevent blockage of downstream larger lines in case it breaks loose.
  — Upstream surcharging should be monitored periodically to avoid basement flooding.

- Weir installation and reading
  — The weir should be tied securely to the manhole steps.
  — The weir should be installed tightly on incoming lines to prevent leakage.
  — The weir should be level and perpendicular to the center of the pipe.

- Readings should be taken not earlier than 20 minutes after installation. If domestic flow is present at the time of reading, a second reading should be taken at least 15 to 20 minutes later to ensure that the minimum flow is actually measured.
- The best results are obtained between 2:00 a.m. and 5:00 a.m.

When O & J readings exclude plugging of upstream segments, upstream flow readings are subtracted from downstream flow readings. Also, weiring time is shorter than with I & M because flow level behind the weir takes less time to stabilize. Otherwise, O & J is similar to I & M.

Equipment care is a consideration. Weirs are not designed to withstand abuse; face plates are made of glass and can break, and the lines and numbers on the bulkhead are plastic laminated but can be scratched off. After usage, they should be cleaned and stored properly. Air plugs should never be over-inflated as they can rupture.

Continuously Recording Measurements. During flow monitoring, each key manhole should be calibrated by measuring velocity and depth for a wide range of flow conditions, including during and after storm events. Key manholes are calibrated by a combination of methods that include dye insertion, velocity probes, and pumping station measurements. Also, in some systems, sanitary bypass structures must be continuously metered during the flow-monitoring period.

DEPTH-VELOCITY METERS. Flow monitoring is best performed with meters capable of simultaneously measuring depth and velocity. An expanding ring, holding a transducer, is inserted to a pipe near the manhole where smooth flow prevails. The meter itself should be secured to a fixed object such as a manhole step (Figure 6.6).

DEPTH-ONLY METERS. Although less expensive than depth-velocity meters, depth-only meters require extensive calibration to determine discharge as a function of depth. The conversion of depth to discharge is the cause of most flow-monitoring inaccuracies. Depth-only monitors are most often calibrated using velocities collected at a particular location. Data are then used to determine the unknown coefficients from either Manning's equation:

$$V = (1/n)\, R^{0.67} S^{0.5} \quad \text{(metric units)}$$

$$V = (1.486/n)\, R^{0.67} S^{0.5} \quad \text{(English units)} \tag{6.7}$$

Where

V = velocity, m/s (ft/sec);
S = slope, m/m (ft/ft);

n = roughness coefficient (at the assumed depth); and
R = hydraulic radius, m (ft).

or the coefficients of some other friction flow equation such as the one proposed by Pomeroy (Metcalf and Eddy, 1981, and Kulin and Compton, 1975):

$$V = K^{1.316} S^{0.54} A_f^{0.316} \tag{6.8}$$

Where

K = roughness coefficient and
A_f = area, m (ft).

The predominant techniques of calibrating a metering location are single-velocity procedures, including

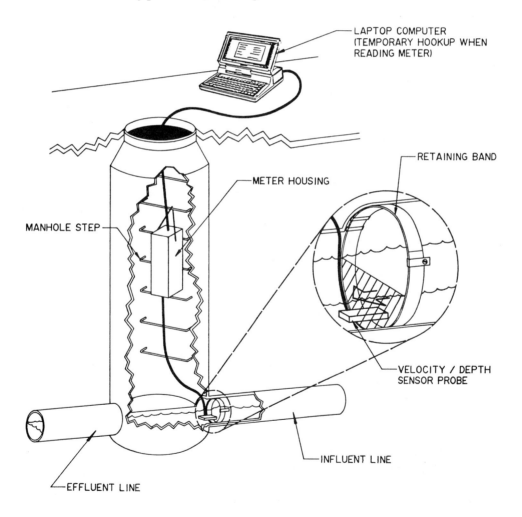

Figure 6.6 Temporary continuous flow recorder installation.

Flow Monitoring

- Assuming average velocity is located 40% of total depth of flow beneath the surface;
- Attempting to find maximum velocity and converting it to average by multiplying by 0.90; and
- The point velocity discharge method.

Single-velocity calibration techniques are unsatisfactory because the equation on which they are based provides a poor fit of sewer velocity profiles. Further, these techniques are error prone because they depend on a single observation of a dynamic situation. Erroneously low velocities resulting from fouling of velocity probes with suspended solids are avoided by a more careful procedure arising from a velocity profile technique. This is confirmed by lower velocities reported in the original, single-velocity technique estimates.

A velocity profile can be developed by using a portable velocity meter. A preliminary reading is taken at the bottom or invert of the pipe, a second reading is collected at a point about midway in the flow, and the third and final reading is taken near the surface. Concurrent manual flow depth readings are also taken.

The flow rate is calculated from the average of the three velocity measurements and the manual depth. A set of velocity and depth readings is obtained from the flow meter immediately before or after obtaining the manual readings, and this information can be used to generate a calibration coefficient for the site. The coefficient is programmed into the meter along with other information, such as the shape and size of the conduit.

Meter sites are visited periodically, and data are collected from each meter electronically with a portable computer. Data present in the meter memory are uploaded to the portable computer, where it can be stored for analysis or copied to floppy disks and downloaded to another computer.

During site visits, batteries in meters should be checked and sensing probes cleaned of accumulated debris.

MEASURING GROUNDWATER. Groundwater levels are good indicators of potential infiltration to sanitary sewers, and should be measured as part of any monitoring program to help correlate and analyze flow.

In certain instances, existing wells in the vicinity may be used to observe groundwater levels. These may be water supply wells, dewatering wells, or cooling water wells, possibly with long-term records of well water levels that can be obtained from the owners.

The number of groundwater gauges used should be a function of factors such as the budget, watershed size and configuration, and soil types in the watershed. There should be enough gauges to allow the engineer to understand the basic groundwater level relationship to sewer elevation. See Chapter 4 for more information on groundwater gauging.

MEASURING RAINFALL. Rainfall measurement devices should be considered an integral part of any sewer system study where I/I information is being analyzed.

Because improper rain gauge location can result in errors in the data, the site should be selected only after consulting a text on hydrology concerning rain gauge installation. The number of gauges required should be a function of watershed size, rainfall variation across the basin, and prevailing storm travel patterns.

Data Analysis

The data collected from flow meters are copied to floppy disks and may be transferred to a desktop computer for analysis. When multiple meter sites are involved, data files should be reviewed and organized according to site number before analysis.

EDITING DATA. Raw data may need editing to make adjustments for flows in meter calibration or when measuring probe installation.

Filtering. Data files can be filtered to eliminate blocks of bad data, such as negative depths or velocities, that could be caused by debris fouling the sensing probe. They can be replaced with values determined by observing typical usage trends on days with similar weather patterns.

Offset. This function adjusts data determined to be inaccurate because a meter is out of calibration by a constant factor.

Time Adjustment. This function may be performed if the clock setting on the metering device is not properly synchronized.

VIEWING DATA. Viewing data allows the analyst to see the trends at the meter site and how they vary based on daily (diurnal) usage or weather-related factors. A graph of flow quantity versus time is known as a hydrograph, which gives the user an immediate impression of activity in a sewer system. It is easier to extract quantified flow values from tabular formats.

DETERMINING KEY VALUES. Determination of key values, such as maximum, minimum, and average daily flows, is facilitated by spreadsheet or database software designed to run on desktop computers. Computer analysis of instantaneous flow rates leads to maximum, minimum, and average flow rates. Dry weather periods are chosen for this part of the analysis to avoid measuring flows contributed by weather-related factors. Analysis to determine I/I rates is discussed more completely in Chapter 4.

QUALITY ASSURANCE

Failure to obtain good data during flow monitoring will jeopardize the success of a rehabilitation program. An integral part of any flow-monitoring project should be a systematic quality assurance program.

MANUAL VERIFICATION OF METER DATA. Automatic electronic flow meters are most frequently selected to perform flow studies. They are subject to drift and other problems that affect calibration over the life of a project, thus periodic manual measurements of depth and velocity are taken with an engineer's ruler and a portable velocity meter. These checks should be made at every meter site each time the meter is serviced. In some cases, slight adjustments will suffice to correct readings, and they can be made during the data analysis phase by comparing meter and manual readings. In other cases, serious problems such as the need for meter replacement can be identified by this process, and spare meters should always be available.

REVIEW OF RECORDED DATA. Field data should be reviewed on a periodic basis, and should be checked for consistency with the prior week's data. Changes should be justified by weather-related or other effects. Waiting until the end of the project to review data does not allow for identification of problem sites or meters. This review should use information obtained from periodic manual measurements.

PROBLEM AREAS. Conditions that can hamper obtaining accurate flow data during a monitoring project include the following:

- The quantity of I/I at specific meter sites may be too small to be detected by a meter;
- Groundwater and weather conditions may not be suitable for recording system responses to wet weather conditions;
- The area monitored may be subject to changing growth patterns; and
- Flow data may have been obtained under surcharge conditions, which prevent an accurate measurement of I/I levels (the project manager must be aware of and account for these possibilities).

REFERENCES

American Society of Civil Engineers (1976) *A Guide for Collection, Analysis and Use of Urban Stormwater Data*. New York, N.Y.

Gutierrez, A.F., and Siu, M. (1982) Flow Measurement in Sewer Lines by the Dye-Dilution Method. Paper presented at the Tex. Sect., Am. Soc. Civ. Eng., Fort Worth, Tex.

Kulin, G., and Compton, P.R. (1975) A Guide to Methods and Standards for the Measurement of Water Flow. NBS Special Publication 421, Inst. for Commerce, Washington, D.C.

Metcalf and Eddy, Inc. (1981) *Wastewater Engineering: Collection and Pumping of Wastewater*. McGraw-Hill, Inc., New York, N.Y.

Water Pollution Control Federation (1983) *Safety and Health in Wastewater Systems*. Manual of Practice No. 1, Washington, D.C.

Chapter 7
Pipeline Rehabilitation Methods

135	General Considerations	151	Contract Documentation
135	Functional Overlap	152	Pipeline Renovation Systems
141	Durability or Life Expectancy	152	Pipe Linings
144	Long-Term Material Properties	152	Sliplining
144	Resistance to Chemical Attack	152	Continuous Pipe
145	Abrasion Resistance	153	Short Pipes
145	Structural Considerations	154	Solid-Wall Polyethylene
145	Loadings	154	Profile Wall Polyethylene
145	Design Types	155	Spiral Rib Polyethylene
145	Hydraulic Capacity	155	Polyvinyl Chloride
147	Cost	155	Fiber-Glass-Reinforced Plastic
147	Systemwide Implications	156	Reinforced Plastic Mortar
148	Construction Issues	156	Cement-Lined or Polyethylene-Lined Ductile Iron
148	Safety		
148	Confined Spaces	156	Steel
148	Trenches	156	Cured-in-Place Pipe
148	Adjacent Structures and Utility Plants	158	Nu-Pipe and U-Liner Deformed Pipe
148	Preparation	159	Rolldown and Swagedown Deformed Pipe
149	Flows		
149	Sewer Cleaning	161	Spiral-Wound Pipe
149	Voids or Loose Ground	161	Segmental Linings
149	Methods of Working	161	Fiber-Glass-Reinforced Cement
149	Materials	162	Fiber-Glass-Reinforced Plastic
150	Pipe Laying	162	Reinforced Plastic Mortar
150	Annular Grouting	162	Polyethylene
150	Supervision	162	Polyvinyl Chloride

Pipeline Rehabilitation Methods

162	Steel	179	Large-Diameter Pipe Grouting
162	Structural and Nonstructural Coatings	181	Materials
		181	Advantages
163	Reinforced Shotcrete	182	Limitations
163	Cast-in-Place Concrete	182	Cost and Feasibility
164	Pipeline Replacement or Renewal	183	External Grouting
165	Trenchless Replacement	183	Chemical Grouts
165	Pipe Bursting	184	Cement Grout
167	Advantages	184	Portland Cement Grout
168	Disadvantages	185	Microfine Cement Grout
169	Microtunneling	185	Compaction Grouting
169	Auger Systems	185	Mechanical Sealing
170	Slurry Systems	185	Point (Spot) Repairs
170	Pipe Installation	187	Manhole Rehabilitation
171	Advantages	187	Manhole Conditions
172	Disadvantages	187	Structural Degradation
172	Other Trenchless Systems	187	Movement and Displacement
172	Directional Drilling	188	Corrosive Environments
172	Fluid Jet Cutting	188	Excessive Infiltration and Inflow
172	Impact Moling	188	Maintenance
172	Impact Ramming	188	Manhole Rehabilitation Methods
173	Auger Boring	189	Chemical Grouting
173	Conventional Replacement	189	Coating Systems
175	Maintenance and Repair	191	Structural Linings
175	Cleaning	193	Corrosion Protection
176	Jet Rodding	193	Frame, Cover, and Chimney Rehabilitation
176	Rodding		
176	Winching or Dragging	194	Service Connection Rehabilitation
176	Cutting	195	Rehabilitation Methods
176	Manual or Mechanical Digging	196	Chemical Grouting
176	Root Control and Removal	196	Cured-in-Place Pipe Lining
178	Pointing	197	Other Measures
178	Internal Grouting	197	References
178	Small- and Medium-Sized Pipes		

Rehabilitating an existing collection system includes upgrading both structural and hydraulic aspects. Structural rehabilitation can include repair and renovation or renewal, while hydraulic rehabilitation can include replacement, reinforcement, flow reduction or attenuation, and (occasionally) renovation. This chapter concentrates on systems that improve the performance of the pipe; those external to the pipeline, such as the provision of storage or the direct removal of input to the network, are not discussed.

Over the past three decades, numerous methods have been developed to repair collection systems. Some of these are restricted in their use to conduits where the existing structure is fundamentally sound or has adequate or spare hydraulic capacity. Others restore the structural capacity of a sewer to that of the original or reduce infiltration and inflow (I/I) to the system, effectively increasing hydraulic capacity. Many of the methods also provide corrosion protection.

Repair methods generally cost less than conventional replacement and involve less open trench excavation, resulting in reduced traffic disruption and public inconvenience. This section addresses established systems with a discussion of some of the more viable new techniques. Inclusion of a particular technique in this manual does not imply endorsement or reduce the responsibility of an engineer in designing and specifying a system.

Where U.S. experience with new materials or techniques is limited, performance records, background information, and case studies from other countries should be considered. The existence of a respected U.S. representative or licensee of such materials or techniques is, however, an important consideration.

In addition to conventional replacement techniques, pipeline rehabilitation systems that have been used with success include renovation systems for pipelines and, in some instances, manholes; minimum excavation techniques; trenchless replacement; and maintenance and repair systems, including cleaning, root removal, pointing, internal and external grouting for pipelines and manholes, mechanical sealing, and point repairs.

The choice of method(s) depends on the physical condition of the sewer system components (pipeline segments, manholes, and service connections) and the nature of the problems. Although this manual is directed at correcting conditions that allow I/I, the rehabilitation methods presented in this chapter also apply to other maintenance, structural deterioration, and hydraulic problems, such as root intrusion, interior corrosion caused by hydrogen sulfide or other chemicals, or lack of capacity.

GENERAL CONSIDERATIONS

The following factors should be considered when selecting a rehabilitation method: functional overlap, durability or life expectancy, structural considerations, hydraulic capacity, cost, systemwide implications, construction issues, and contract documentation.

FUNCTIONAL OVERLAP. The performance of a drainage system can be assessed in two categories: hydraulic and structural. Depending on whether it is a sanitary, stormwater, or combined system being evaluated, the hydraulic category can be divided into public health or flooding and environmental impact. Because solutions to either of these performance areas can have a major impact on the other, rehabilitation of a network should be investigated in a coordinated manner over a complete drainage basin or subbasin.

Investigation procedures and their interdependence are discussed in Chapters 3 and 4. Various rehabilitation techniques are suited to solving particular performance problems and could enhance or adversely affect others. The principal advantages and disadvantages and potential applications of pipeline rehabilitation methods are shown in Tables 7.1 to 7.3; of manhole

Table 7.1 Pipeline rehabilitation options—renovation.

Rehabilitation option[a]	Principal advantages	Principal disadvantages	Potential applications
Pipe linings			
Sliplining			
Continuous pipe (Fusion-welded PE/PB)	Quick insertion Large-radius bends accommodated	Circular cross section only Insertion trench disruptive High loss of area in smaller sizes Less cost effective where deep	4 to 63 in.[b]
Short pipes (PE, PB, PVC, RPM, FRP, DI, ST)	High strength-to-weight ratio Variety of cross sections can be manufactured Minimal disruption	Some materials easily damaged during installation Larger pipes may require temporary support during grouting May involve labor-intensive jointing	4 to 144 in.
Cured-in-place pipe (CIPP)	Rapid installation No excavation Accommodates bends and minor deformation Maximizes capacity Grouting not normally necessary	Full bypass pumping necessary Sole source often necessary High set-up costs on small projects	4 to 108 in.
Deformed pipe			
U-Liner/Nu-Pipe	Rapid installation Continuous pipes Maximizes capacity No excavation Grouting not required	Lateral relocation may be difficult Relies on existing pipe for support	2.5 to 24 in.

Table 7.1 Pipeline rehabilitation options—renovation (continued).

Rehabilitation option[a]	Principal advantages	Principal disadvantages	Potential applications
Swage lining/roll down	Rapid installation Maximizes capacity Minimal excavation Grouting not required	Lateral relocation may be difficult Relies on existing pipe for support	3 to 24 in.
Spiral-wound Pipe	Tailor-made inside the conduit No excavation required Maximizes capacity Rapid installation Noncircular available	Large number of joints Relies on existing pipe for support Requires careful grouting of annulus	3 to 120 in.
Segmental linings FRC, FRP, Gunite	High strength-to-weight ratio Variety of cross sections can be manufactured Minimal disruption	Some materials easily damaged during installation May require temporary support during grouting Labor intensive Requires person entry	36 in. and larger
Coatings Gunite/shotcrete	Connections easily accommodated Zero/minimal excavation Variety of cross sections possible	Difficult to supervise May be labor intensive Control of infiltration required	4 ft[c] and larger

[a] PE = polyethylene, PB = polybutylene, PVC = polyvinyl chloride, RPM = reinforced plastic matrix, FRP = fiber-reinforced plastic, DI = ductile iron, ST = steel, and FRC = fiber-reinforced cement.
[b] in. × 25.40 = mm.
[c] ft × 0.304 8 = m.

Table 7.2 Pipeline rehabilitation options—minimum excavation and trenchless replacement.

Rehabilitation option	Principal advantages	Principal disadvantages	Potential applications
Minimum excavation Narrow trenching	Reduced surface disruption Good size availability	Potential damage to services Off-line only	
Trenchless replacement			
Pipe bursting	Can replace a variety of materials Size for size or size increases Not dependent on condition of existing conduit	Potential damage to adjacent services Lateral connections require disconnection Full bypass pumping required Only suitable for brittle pipes	4 to 20 in.[a] existing
Microtunneling (includes pipejacking)	High groundwater heads Slurry can be water Can deal with cobbles Small-diameter shafts Can excavate plain, weak concrete	Service connections Bentonite slurry requires treatment Off-line only	6 to 36 in.
Directional drilling	Rapid installation Long distances Can be used in tidal or surf zone and under water Variety of pipe materials	Service disruption Generally not suitable for gravity lines Difficult to use in sandy/granular material Off-line only	4 to 36 in.
Fluid jet cutting	Range of up to 400 ft[b] Accurate steering Capable of steering around obstructions Minimal surface disruption Small, self-contained equipment	Possibility of service damage Operation difficult in sandy or granular soils Not suitable for gravity lines Off-line only	2 to 14 in.

[a] in. × 25.40 = mm.
[b] ft × 0.304 8 = m.

Table 7.3 Pipeline rehabilitation options—conventional replacement and maintenance or repair.

Rehabilitation option	Principal advantages	Principal disadvantages	Potential application
Conventional replacement			
Open cut	Removes all problems on length Traditional design	Expensive, particularly if deep Disruptive	Any
Tunneling	Removes all problems on length Traditional design Reduces disruption Flexibility on line/elevation	Usually more expensive than open cut May need expensive ancillary works	Greater than 3 ft[a]
Maintenance and repair			
Cleaning	Increases effective capacity May resolve localized problems	May be costly and cause damage May become a routine requirement	Any
Root removal	May increase effective capacity May resolve localized problems	May be costly Problem likely to recur	Any
Pointing	Restores original condition cheaply Minimal disruption Increases capacity	Person-entry sewers only Sewer must be structurally sound	Greater than 3 ft
Internal grouting	Seals leaking joints and minor cracks Prevents soil loss Low cost and causes minimal disruption Can reduce infiltration Can include root inhibitor	Infiltration may find other routes of entry Existing sewer must be structurally sound May recur/become routine requirement	Any

Pipeline Rehabilitation Methods

Table 7.3 Pipeline rehabilitation options—conventional replacement and maintenance or repair (continued).

Rehabilitation option	Principal advantages	Principal disadvantages	Potential application
External grouting	Improves soil conditions surrounding conduit Can reduce infiltration and soil loss	Difficult to assess effectiveness Can be costly May recur/become routine requirement	Any
Mechanical sealing	Seals leaking joints and minor cracks Prevents soil loss Low cost and causes minimal disruption Can reduce infiltration Can include root inhibitor	Infiltration may find other routes of entry Existing sewer must be structurally sound Suitable for person-entry sewers only	Person-entry only
Spot repairs	Deals with isolated problems	Requires excavation for small conduits May require extensive work on brick sewers	Any

[a] ft × 0.304 8 = m.

rehabilitation methods are shown in Table 7.4; and of service connection rehabilitation methods are shown in Table 7.5.

Proper selection of rehabilitation methods and materials depends on a complete understanding of the problems and other contributing factors. Decisions should include an evaluation of both internal conditions, based on some form of visual inspection, and conditions surrounding the pipe (see Chapter 3).

DURABILITY OR LIFE EXPECTANCY. Engineers may be reluctant to recommend newer technology over traditional methods because of uncertainties over the life expectancy or durability of the technology. Uncertainties often are reinforced by well-publicized failures of newer materials in other fields of construction. Many failures occur because of poor design and construction practices and insufficient knowledge of the limitations of the material.

Not all new rehabilitation technologies use new and untried materials; some simply use a new method of installing traditional materials or use established materials in new applications. An example is microtunneling, a new method of installing traditional concrete pipes with minimal surface disruption.

Another factor that discourages the use of modern materials and technology is the historically demonstrated durability of traditional materials. Unless installed in a particularly harsh environment, the life expectancy of a conduit constructed of traditional materials can be more than 100 years, such as brick or vitrified clay sewers constructed during the 1800s that show few signs of deterioration.

In contrast, the long-term characteristics of a newer material must be estimated by identifying conditions that may degrade its performance, both generally and specifically, in the sewer environment. Accelerated aging tests are then developed by increasing the concentration of an aggressive chemical or exposing the material to elevated temperatures. Both traditional and new materials are included in the tests to provide a benchmark against which data can be compared.

For more established materials, enough data have been accumulated in the U.S. and other countries to predict design lives of 50 to 100 years, assuming materials have been manufactured and tested to an adequate specification, design procedures are properly followed, and advice about exposure to adverse conditions is heeded. Data from other countries should be translated carefully because operating conditions, specifications, testing procedures, manufacturing standards and materials, and initial and long-term properties may be different. Factors affecting durability include long-term material properties and resistance to chemical attack and abrasion.

Table 7.4   Manhole rehabilitation options.

Rehabilitation option	Principal advantages	Principal disadvantages
Rehabilitation of manhole structure by plugging, patching, and coating with mortars, coatings, and sealants (both cementitious and noncementitious).	Improve structural condition, eliminate leakage, and provide corrosion protection. Little disruption.	Will not rehabilitate badly deteriorated or structurally unsound manholes.
Repair or rebuilding of manhole chimney and cone section when excavation is required.	Rehabilitate badly deteriorated or structurally unsound chimney and cone section.	Evacuation required.
Step removal or replacement.	Improve access and safety and eliminate leakage.	Installation difficulty, cost.
Replacement of manhole frame and cover.	Improve service life and alignment, adjust grade, and eliminate leakage.	Excavation required, cost.
Structural relining (linings and gunite).	Renew structural integrity.	Reduction of diameter, cost.
Sealing of frame-chimney joint and chimney above cone.	Eliminates inflow while allowing movement of the frame.	Minor reduced access in chimney, cost.
Seal or replace cover or install insert.	Eliminates inflow and stops rattle.	Raises cover slightly.
Chemical grouting of manhole structure.	Eliminates infiltration and fills voids in surrounding soil.	Does not improve or rehabilitate interior of manhole.
Total replacement.	New manhole.	Cost.

Table 7.5 Service connection rehabilitation options.

Rehabilitation option	Principal advantages	Principal disadvantages	Potential applications
Chemical grouting	Seals leaking joints and minor cracks Prevents soil loss Low cost and minimal disruption Can reduce infiltration Can include root inhibitor	Infiltration may find other routes of entry Existing sewer must be structurally sound May recur/become routine requirement	4 in. and larger[a]
Inversion lining	Rapid installation No excavation Accommodates bends and minor deformation Maximizes capacity Grouting not normally necessary	Full bypass pumping necessary High set-up costs on small projects	6 to 42 in.
U-Liner/Nu-Pipe	Rapid installation Continuous pipes Maximizes capacity No excavation Grouting not required	Lateral relocation may be difficult Relies on existing pipe for support	2.5 to 24 in.
Swage lining/Roll down	Rapid installation Maximizes capacity Minimal excavation Grouting not required	Lateral relocation may be difficult Relies on existing pipe for support	3 to 24 in.
Spiral-wound pipe	Tailor-made inside the conduit No excavation required Maximizes capacity Rapid installation Noncircular available	Large number of joints Relies on existing pipe for support Requires careful grouting of annulus	3 to 120 in.

[a] in. × 25.40 = mm.

Pipeline Rehabilitation Methods

Long-Term Material Properties. Degradation may occur through long-term changes in the nature of the material that may or may not be linked to the specific environment. For example, the strength and stiffness of plastic materials are stress and time dependent. The result is that materials progressively deform when subjected to load, a condition known as creep. The rate at which this occurs varies from plastic to plastic and often causes reduced strength and stiffness, which is not necessarily a problem if the effects are considered in design calculations.

Chemical or microbiological factors also may affect long-term properties of polymeric materials. Potential problems often are realized by manufacturers, and many established materials are well-engineered for the sewer environment.

Resistance to Chemical Attack. The resistance of established materials to a wide range of potentially aggressive agents has been researched and documented. A general comparison of the various materials is given in Table 7.6. Manufacturers should be able to provide more detailed information regarding the resistance of their materials to specific chemicals or chemical combinations, though it is impossible to guarantee adequate protection against accidental discharges of large quantities of corrosive materials.

Table 7.6 General comparison of durability of pipeline.

Type of material	Durability		
	Abrasion resistance	Chemical and corrosion resistance	Integrity of jointing
Vitrified clay pipe	Medium	High	Medium
Ductile iron/steel	High	Low*	Medium
Asbestos cement	Low	Low*	Medium
Concrete	Low	Low*	Medium
Fiberglass-reinforced concrete	Low	Low*	Medium
Gunite/shotcrete	Low	Low*	Low
Reinforced plastic mortar/fiber-reinforced plastic	Medium	Medium*	Medium*
Cured-in-place pipe	Medium*	Medium*	N/A
Polyethylene			
Continuously welded mechanical joints	High	High	Medium–high

* Insufficient data to distinguish between higher and lower value.

Abrasion Resistance. Polymeric materials generally are more resistant to abrasion than cementitious materials, particularly where the effluent is of low pH. Polyethylene (PE) and polyurethane are most resistant to abrasion, but fiber-reinforced plastics can present problems because the integrity of the thin surface liner can significantly affect long-term structural properties.

STRUCTURAL CONSIDERATIONS. All rehabilitation systems must be designed to withstand short-term installation and long-term loading conditions.

Loadings. In the case of linings, short-term loadings include longitudinal winch or jacking forces and flotation and buckling forces exerted by grouting. Flotation and buckling forces cause small, permanent deformations that must be considered when determining long-term structural performance. Long-term loadings (ground, traffic, or surcharge loading) should be able to withstand pressures imposed by both internal surcharge, which depends on the operation of the system, and external groundwater conditions.

Design Types. Linings can be divided into three basic long-term design types:

- Composite design is used where there is a bond between the lining, annulus grout, and existing sewer; all three act as a composite structure.
- Flexible design is used for pipe linings where no reliable bond exists, and the structural improvement is provided by the lining. The grout and existing sewer provide the support necessary for what is essentially a flexible pipe within the existing pipe.
- Rigid design is used where the lining pipe has sufficient rigidity to act as an independent structure. Any support provided by the grout and existing sewer enhances the load-bearing capacity.

All the linings discussed in the following sections can be considered in one of these categories. The trenchless replacement technology can be designed using conventional rigid pipe, flexible pipe, or tunnel design theories.

HYDRAULIC CAPACITY. One consideration in evaluating pipeline rehabilitation alternatives is the effect of the sewer system on flow capacity. In determining the flow capacity of pipelines, contributing factors include flow area, pipe shape (for example, circular, elliptical, or arch), slope and internal roughness, nature of pipe flow, and downstream restrictions. Most pipeline rehabilitation alternatives change the flow area and internal pipe roughness, and sometimes pipe shape.

Changes in pipe size or shape associated with pipe rehabilitation typically reduce the flow area and hydraulic radius components of pipe flow calculations. These reductions normally are smaller with tight-fitting pipe rehabilitation products that conform to the walls of the original pipe. Reduction can be more substantial where standard circular pipe sections are sliplined within the old pipe, forming an annulus between the two pipes.

Despite reductions in flow area, available pipe capacity often can be maintained or increased because of reductions in I/I entering the system and improvements in existing internal pipe roughness. Although reductions in I/I vary considerably for each system and are difficult to quantify, any reduction of extraneous flows can provide more pipe capacity for sanitary wastewater.

Pipe flow capacity typically is enhanced through decreases in internal pipe roughness, which becomes apparent when evaluating the following factors affecting pipe roughness determined through research conducted by the Water Research Centre (WRC) in the U.K. (*Sewerage Rehabilitation Manual*, 1986): roughness of the pipe surface (based on the pipe material), pipe joint eccentricity or irregularity, aging of the pipe material, sliming, large-scale roughness (caused by pipe surface irregularities and solids deposits), and sediment.

Many of these factors improve over the long term through pipeline rehabilitation. Water Research Centre testing indicated that plastic pipe materials shed slime more readily than more porous concrete and clay materials, resulting in less slime accumulation over time, which decreases pipe roughness. Additional data indicate that the pipes roughest when new continue to be the roughest when fully slimed (*Manual on Buried*, 1982).

Many pipeline rehabilitation options involve plastic pipe materials, which are smoother when new than traditional pipeline materials. Unlike concrete pipe and cement mortar in brickwork sewers, plastics do not roughen with age.

Joint and surface irregularities (large-scale roughness) are reduced or eliminated with many pipeline rehabilitation alternatives. With form-fitting products, abrupt and rugged irregularities in the underlying pipe are replaced by smooth transitions. In conventional sliplining with a smaller diameter pipe, some irregularities can be eliminated. Joints and cracking in the original pipe also reduce flow capacity by providing access for roots and sediment to enter the pipe. With jointless pipeline rehabilitation products, these points of access are eliminated, which will significantly improve pipe roughness over the long term.

All of these factors combine to indicate enhanced, long-term pipe roughness characteristics of continuous pipe rehabilitation products. Testing in the U.S. on in-service sanitary sewers supports this by comparing pipeline rehabilitation products to original, deteriorated pipeline materials.

A study was conducted to evaluate the comparative pipe roughness of uncleaned, in-service Insitupipe™ and more traditional pipeline materials in sanitary sewers (Sewer Segment, 1990). A dye-dilution technique was used to measure the flow rate in domestic wastewater sewers with a diameter of 375- to 1 350-mm (15- to 54-in.). Manning *n* roughness coefficients were found to be 33% smoother for in-service Insitupipe™ sewers versus original in-service traditional clay and concrete sewers.

There is some justification in recommending the use of lower Manning n values for lined sewers versus those used for the deteriorated existing pipeline, but further work is needed before extending the comparison to cover complete replacement with nonplastic materials. In the interim, all new pipelines should be designed using the same long-term n value, regardless of pipe material.

COST. The initial costs of different rehabilitation methods traditionally have been approximately the same because all methods have similar long-term effectiveness and service lives. Most aspects of conventional sewerage systems are designed with the assumption they will have a long life; that they may have to be renewed, assuming they are still needed, is understood. Cost appraisals should account for this, particularly when alternatives with different life expectancies are considered. Because life expectancies of pipeline assets are comparatively long, such appraisals can become academic.

Cost appraisals are necessary where shorter lives can be expected or newer rehabilitation techniques with limited experience are involved. When comparing a technique with a life expectancy of 50% of the alternative, an allowance should be included for costs likely to occur when current techniques reach the ends of their useful lives. A perceived initial saving may ultimately cost more.

One principal aim of rehabilitation is to use the existing sewer system, with associated savings in allied costs. These can include the cost of disruption caused by construction (for example, traffic delays and diversions, loss of trade and access restrictions to businesses, noise, dirt, and health hazards) and the effects on other utility equipment. These costs should be considered when evaluating alternatives.

SYSTEMWIDE IMPLICATIONS. Any rehabilitation option will affect the performance of other parts of the system. For example, renovation or replacement will affect hydraulic performance by increasing or decreasing pipe flow, potentially increasing surcharging and flooding and increasing pollution caused by more frequent spills from storm or wastewater overflows. Such problems may not be limited to the pipe in question.

Similarly, solutions to hydraulic problems may result in more frequent surcharge in structurally defective pipes, increasing the likelihood of failure. Thus the systemwide effects of a proposed solution should be considered so that other problems are not created. In larger systems, this can be achieved by analyzing the proposed solution in a hydraulic model of the system.

CONSTRUCTION ISSUES. As with all construction projects, rehabilitation systems are affected by many external factors. Though most manufacturers and specialty contractors have considered these problems when designing their products, when selecting a technique the engineer should study construction-related issues such as safety, preparation, methods of working, and supervision.

Safety. The most important construction-related issue is safety. The construction manager, contractor, and the contractor's employees must perform the construction safely. This relates not only to the contractor's employees but also to third parties, adjacent structures and the utility plant.

CONFINED SPACES. Because most rehabilitation systems incorporate the existing pipeline or connect to an existing sewer network, the construction contract will involve entry into confined spaces. In such circumstances, the confined atmosphere must be monitored before and during confined space entry, and safe access and working procedures must be followed. The specifying agency must inform the contractor that such conditions will exist and are to be included in the contract price.

TRENCHES. Although many systems are described as trenchless, additional accesses often are needed to install the pipe or remake connections, in which case the contractor must ensure that all excavations meet appropriate standards for trench support and safety.

ADJACENT STRUCTURES AND UTILITY PLANTS. A major advantage of trenchless operations is that they limit surface disruption in congested areas such as busy streets, narrow alleys between buildings, and commercial and industrial centers. Construction in such locations is more likely to damage adjacent structures and utility plants. Trenchless systems also can cause disruption, though, as in pipe bursting methods damaging adjacent pipes if the bursting becomes too close, the degree of upsizing is too great, or ground conditions magnify vibrations. The designer, construction manager, and contractor all should be aware of the potential for such problems and mitigate them through adequate design.

Preparation. When rehabilitating an existing pipeline or using trenchless technology, some disruption is inevitable and appropriate arrangements should be made during contract bidding. Existing accesses could be used, which may require removing the cover and frame and modifying the access structure. Where access does not meet current standards, it should be improved, which is particularly important because I/I flow could still enter through the manhole after pipeline rehabilitation. The spacing of access shafts will affect installation costs, which must be balanced with the costs of providing additional access.

The size of a lining depends on the minimum dimensions of the existing conduit. Often it is impossible to check the size before the contract begins because of sediment, high flows, and lack of access. Full-scale models or sections of pipe should be used to prove a sewer is essential before the lining pipe is ordered.

Some excavation will be necessary for such items as lead-in trenches for sliplining and connections. Knowledge of other utility lines and structures is important.

FLOWS. It generally is necessary to divert all flows from a sewer during installation, which can produce a major portion of rehabilitation costs. However, some techniques are helped by the presence of low flows. An understanding of sewerage system operations is important to assessing the required diversion capacity required or the rapid change during wet weather, tides, and industrial discharges.

High infiltration levels may require special attention be given to jointing methods, annulus grouting, pointing, and coatings so that liner integrity is not compromised.

SEWER CLEANING. Precleaning existing sewers permits the entry of linings and promotes the bond between the lining material and sewer. Sediment and debris in the invert of the sewer can also harm some lining materials, leading to early failure. Jet rodding often can remove smaller material, although larger debris may require mechanical removal.

Severe encrustation and scale on the sewer walls may prevent some rehabilitation systems from being successful. Joint sealing equipment may not seal effectively, and sliplining may require an unacceptable reduced size.

The sewer should be inspected after cleaning and immediately before rehabilitation to ensure there are no obstructions that could inhibit lining. Flows should be removed or significantly reduced during inspection.

VOIDS OR LOOSE GROUND. Where significant voids are known or suspected to exist, secondary grouting should provide appropriate pipe support. Such grouting is unlikely to be satisfactory in the long term when used alone to reduce infiltration.

Methods of Working. *MATERIALS.* Materials should be stored and handled to prevent damage. Polyethylene materials are prone to creep and should be stored and transported correctly. Manufacturers' guidelines for the height of pipe stacks should be followed. Liner pipes are available in a variety of lengths, the specific length for a given project being determined by handling or construction considerations. More detailed materials recommendations are discussed in Chapter 8.

PIPE LAYING. Contrary to conventional pipe-laying practice, linings normally are installed downstream to assist installation and reduce the potential for fouling the clean annular space with wastewater. Annular space that could be fouled should be grouted.

ANNULAR GROUTING. There are a variety of methods for grouting the annulus, depending on the size of the sewer to be renovated. Annular grouting will result in flotation and buckling forces being applied to the lining, though the risk of buckling could be reduced by filling the lining with pressurized water. Unless the lining is supported at relatively close intervals, single-stage grouting could cause lining flotation. Staged grouting, a technique frequently used in larger systems, considerably reduces the grouting forces and may eliminate or reduce flotation.

Two types of annular grouting methods exist, related directly to the availability of access to the inside of the liner. While the methods vary, several general rules can be considered when grouting (Lee, 1991). The volume of grout required should be estimated and infiltration to the existing sewer limited so the grout is not diluted. The grout should be injected from the downstream end and from the invert to the crown, with air vents provided at the high points. Grout pressures at the nozzle, typically 34.5 kPa (5 psi) or approximately one-third the pipe stiffness, should be monitored throughout, and the volumes of grout used should be measured and compared with the original estimate. Grout volumes used often differ considerably from those calculated.

Supervision. Because a main cause of failure in sewer rehabilitation is poor workmanship, it is cost effective to emphasize workmanship standards and increased site supervision. The many underground operations involved necessitate the development and use of appropriate inspection procedures and controls. This is particularly important for systems heavily dependent on in-place curing for their performance. In small-diameter sewers, where access is limited or impossible, closed-circuit television (CCTV) equipment can be valuable.

Annular grouting needs close supervision, as accurate measurement of grout quantities indicates leakage, the presence of voids, or difficulties in the grouting method. Line, level, and deformation of the lined sewer also should be monitored. Typical grout tests include workability, density, bleed, setting time, and compressive strength. There is currently no reliable and inexpensive method of checking grout penetration and fill. In larger sewers, the lining can be tapped with a hammer to detect hollow, ungrouted areas, although unbonded linings cannot be distinguished from voids. Small 8-mm (0.25-in.)

holes can be drilled to confirm the presence of grout, and 50-mm (2-in.) cores taken where there is sufficient access. Unless holes are carefully repaired, they can provide a weakness in the newly rehabilitated pipeline.

CONTRACT DOCUMENTATION. Many rehabilitation techniques are either new systems or new uses for traditional materials, thus standard specifications often are not available. Even where such documentation is available for existing materials, new uses may not be covered. For example, while specifications and codes of practice have long been available for sliplining sewers with high-density polyethylene (HDPE), they do not cover recent developments in deformed pipe lining. Therefore, engineers must carefully consider the potential impacts of new installation practices on material durability and must specify materials appropriately. Advice from qualified materials specialists may help in understanding implications and specification needs.

Because techniques are constantly developing, new or improved materials may not meet some of the old specification requirements. This is particularly important for material typing tests, where the material is checked for both short- and long-term suitability in the proposed environment. Where standard specifications are not available, documents should be found that cover areas traditionally covered by referring to other standards. In the U.K., interim specifications cover newer rehabilitation options (Information and Guidance, 1986). These documents include sections on the following:

- Material-type tests to prove a particular material suitable and to provide the material properties required for design;
- Product-type tests to prove the material can be manufactured in the size and shape unit required;
- Quality control tests to ensure all units are manufactured to the same quality as those originally approved under the type tests;
- Definitions of terms where none exist or where existing ones are deficient;
- Dimensional tolerances giving the contractor, manufacturer, and construction manager a measure of material quality;
- Workmanship, inspection, and certification to ensure the pipeline is installed and manufactured according to specifications;
- Jointing systems, particularly where these have a significant impact on structural integrity or where various types are available; and
- Pipe construction and manufacture (because materials have various types of construction, some suitable and some unsuitable for rehabilitation, the particular type of material construction should be specified).

Manufacturers and contractors often can offer advice in many of these areas, but the engineer still must ensure the specifications are correct.

*P*IPELINE RENOVATION SYSTEMS

Pipeline renovation systems use the existing structure either to form part of the new pipeline or to support a new lining. Such systems can be divided into three categories: pipe linings, segmental linings, and structural and nonstructural coatings.

PIPE LININGS. Pipe linings may be installed inside the sewer in either continuous or short lengths. A large variety of sewer sizes and cross sections can be lined, using sliplining (continuous pipe and short pipes), cured-in-place pipe (CIPP), deformed pipe, or spiral-wound pipe.

Sliplining. Sliplining is a method by which pipes are inserted to an existing line by pulling or pushing continuous or short-length pipes into the sewer. The annulus between the existing pipe and liner pipe should be grouted. There are many pipe types available.

CONTINUOUS PIPE. Sewers may be sliplined with continuous lengths of solid-wall PE or polybutylene pipes, commonly supplied in 12-m (40-ft) lengths and butt fusion-welded together before insertion to the sewer pipe. Pipes are drawn or jacked into the sewer, subsequent lengths being added as required. Where the annulus between the lining and the existing sewer is 25 mm (1 in.) or more, it should be grouted to provide support for the flexible lining. Where the annulus is not grouted, the liner should not be considered a structural lining. Small-diameter, continuous, extruded pipe can be used. A nose cone is attached to the front of the pipe to reduce the likelihood of snagging and as a point for attaching the winch cable.

For pipes less than approximately 600 mm (24 in.) in diameter, the insertion time can be reduced by prewelding the pipes on the ground surface before installation. The pipes are then inserted through a lead-in trench (Figure 7.1), the size of which depends on the size of the lining pipe and the depth of the sewer. Larger, thicker walled, or deep pipes normally are welded in the bottom of the insertion trench, reducing the size of the excavation and eliminating the need to lay the pipe string out on the surface (Figure 7.2).

Several hundred feet can be sliplined in one operation, with pipe flexibility allowing negotiation of large-radius bends. Flows should be controlled to keep rags and solids from lodging in the annulus, which would be detrimental to high-confidence sliplining or grouting.

Cleaning the pipeline by jetting or mechanical means will remove roots, encrustation, and sediment to facilitate lining insertion. If hydraulic capacity is to be increased, intruding connections must be removed and point (spot) repairs may be necessary in some locations before sliplining.

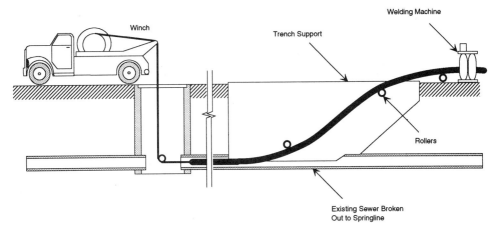

Figure 7.1 Traditional sliplining.

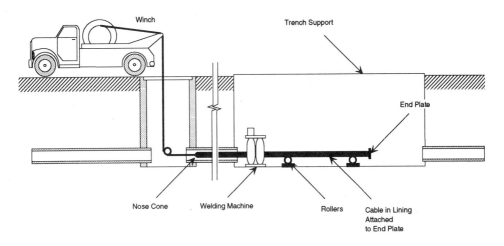

Figure 7.2 Welding in trench.

Polyethylene is a flexible pipe requiring support from surrounding material to ensure its long-term stability. It does not bond with the grout; thus sliplining with solid-wall PE should be designed using the flexible design theory. Side sewers, service connections, and laterals typically are reconnected by excavation from the surface.

SHORT PIPES. As an alternative to sliplining with continuous pipes, short pipe lengths can be joined by gasketed or mechanical joints. They are inserted either by carrying individual pipes into the sewer and jointing them *in situ* or by jacking pipes into place and jointing them at the insertion pit (Figure 7.3). *In situ* jointing is suitable only for larger sewers (greater than

Pipeline Rehabilitation Methods

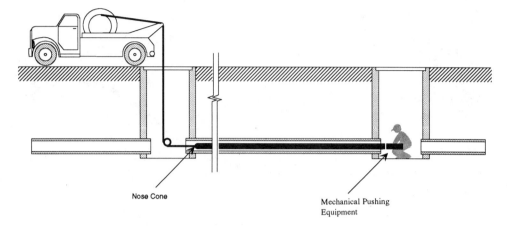

Figure 7.3 Insertion method for sliplining with short pipes.

1 050 mm [42-in.] in diameter) but provides greater flexibility for negotiating bends because special pipes can be manufactured and installed. Short pipes are manufactured in a wide variety of sizes and shapes from a number of materials.

Solid-Wall Polyethylene. Short lengths of solid-wall PE pipes are available with machined screw or snap-fit joints for insertion in smaller diameter sewers. The design of the joint permits constant internal and external diameters. Joints must be watertight and not pull apart during installation. Pipes are available in sizes up to 1 575 mm (63 in.) in outside diameter and can be inserted from relatively small pits, of a length dependent on the length of pipe chosen. Linings sometimes can be installed through existing accesses.

Profile Wall Polyethylene. Profile wall PE is manufactured in 6-m (20-ft) lengths and is connected in a bell-and-spigot joint configuration with an elastomeric sealing gasket. The inner profile wall design permits constant inside and outside diameters. A jacking pit is excavated approximately 1.5 m (5 ft) longer than the inner profile wall pipe segment length and 0.6 m (2 ft) wider than the existing sewer pipe outside diameter. Insertion pit dimensions normally are dictated by the dimensions of the jacking equipment. The top half of the old pipe in the pit is removed for liner pipe access, and the pipe is pushed or pulled through the existing pipe.

After insertion, the annular space between the pipes is pressure grouted. Pressures of approximately 34.5 kPa (5 psi), or approximately one-third of the pipe stiffness, should be used to prevent collapse of the liner pipe during grouting. Jacking can be accomplished with some flows in the sewer, and the liner inserted upstream or downstream. Service connections or laterals are remade using procedures similar to those for PE continuous pipe.

Pipes are available in diameters ranging from 450 to 3 600 mm (18 to 144 in.) and have been inserted more than 300 m (1 000 ft) from one insertion pit. Short sections have been pushed through curves while maintaining normal sewer flows.

Inner profile is a flexible pipe requiring support from the surrounding material to ensure its long-term stability. As it does not bond with the grout, sliplining with profile PE should be designed using the flexible design theory.

Spiral Rib Polyethylene. This PE pipe is manufactured in 6-m (20-ft) lengths and connected in a bell-and-spigot joint configuration with an elastomeric sealing gasket. The spiral rib design permits a constant inside diameter with a varying outside profile rib, depending on annular space and liner pipe design requirements. The jacking pit and other insertion procedures are similar to those for profile wall PE pipe. Pipes are available in sizes from 455 to 3 660 mm (18 to 144 in.) in diameter.

Spiral rib PE is a flexible pipe requiring support from the surrounding material for its long-term stability. It does not bond with the grout, so sliplining with spiral rib PE should be designed using the flexible design theory.

Polyvinyl Chloride. Polyvinyl chloride (PVC) is manufactured in 3- or 6-m (10- or 20-ft) lengths. It also has been used in 1-m (3-ft) lengths, which are inserted from within a manhole. Lengths are connected in a bell-and-spigot configuration with an elastomeric sealing gasket. A solvent cement is used in sewer bypass conditions. The jacking pit and other procedures are similar to those for profile wall PE pipe. Pipes are available in diameters ranging from 100 to 900 mm (4 to 36 in.) and are only available in circular cross sections.

Polyvinyl chloride is a flexible pipe requiring support from the surrounding material for long-term stability. It does not bond with the grout, so sliplining with PVC should be designed using the flexible design theory.

Fiber-Glass-Reinforced Plastic. Fiber-glass-reinforced plastic (FRP) pipe usually is manufactured in 12-m (40-ft) lengths, but 6- and 18-m (20- and 60-ft) lengths also are available. Lengths are connected in a bell-and-spigot configuration with a fiber glass, overwrap, sealing closure. Gasketed and cement joints also are available. The filament-winding manufacturing process provides a constant inside diameter. Outside diameter typically is controlled by the bell size, which is larger than the barrel size. The jacking pit and other procedures are similar to those for profile wall PE pipe.

Available in inside diameters ranging from 100 to 4 225 mm (4 to 169 in.), the pipe can accommodate bends and is suitable for a variety of cross sections.

Fiber-glass-reinforced pipe is flexible and requires support from the surrounding material to ensure its long-term stability. As it does not bond with the grout, sliplining with FRP should be designed using the flexible design theory.

Reinforced Plastic Mortar. Reinforced plastic mortar (RPM) pipe (also known as reinforced thermosetting resin, or RTRP) is manufactured in 6-m (20-ft) lengths, connected with a bell-and-spigot joint configuration and an elastomeric sealing gasket. The centrifugal manufacturing process provides a constant outside diameter with a slightly larger dimension at the joint. When thicker-walled pipe is used, the joints can be recessed so the outside diameter is constant. The jacking pit and other procedures are similar to those for profile wall PE pipe. The pipe is available in diameters ranging from 450 to 2 400 mm (18 to 96 in.). This pipe cannot easily accommodate bends and is only available in circular cross sections.

Reinforced plastic mortar is a flexible pipe requiring support from the surrounding material to ensure its long-term stability. It does not bond with the grout, and sliplining with RPM should be designed using the flexible design theory.

Cement-Lined or Polyethylene-Lined Ductile Iron. Ductile iron pipe is manufactured in 6-m (20-ft) lengths connected in a bell-and-spigot joint configuration with an elastomeric sealing gasket. The centrifugal manufacturing process provides a constant outside diameter. However, the bell is several inches larger and controls the size for insertion. The jacking pit and other procedures are similar to those for profile wall PE pipe. The pipe, which cannot easily negotiate bends, is available in diameters ranging from 100 to 1 500 mm (4 to 60 in.). Ductile iron is a semirigid material and should be designed as a semirigid pipe.

Steel. Steel pipe is manufactured in 12-m (40-ft) lengths, although 6-m (20-ft) lengths also are available. Lengths are connected by welding the plain ends. The welded joint closures necessitate bypassing wastewater flows during construction. Joint welding can be completed within the insertion pit area. The jacking pit and most of the other procedures are similar to those for profile wall PE pipe. The pipe is available in diameters ranging from 100 to 3 600 mm (4 to 144 in.). This pipe cannot easily negotiate bends. Steel is a flexible material and should be designed as a flexible pipe.

CURED-IN-PLACE PIPE. The CIPP lining process involves inserting a flexible lining to the sewer using either an inversion process (Figure 7.4) or a winching process that is now well established in the U.S. The lining is inserted through existing manholes and, depending on the system selected, installed using water inversion, winched inversion, or air inversion. With water inversion, the lining is inverted under water pressure and cured by circulating hot water. With winched inversion, the lining is winched into place and in-

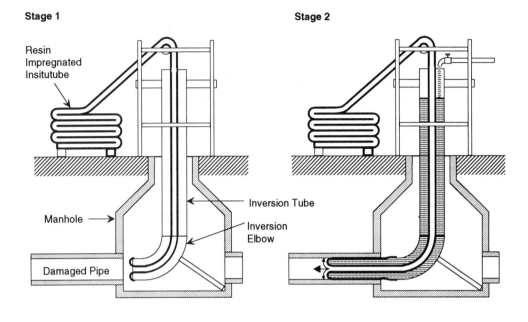
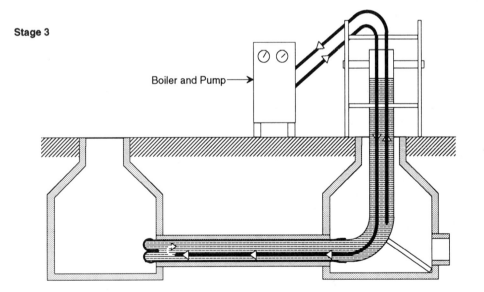

Courtesy of Insituform™ of North America

Figure 7.4 Cured-in-place pipe installation.

Pipeline Rehabilitation Methods 157

flated against the sewer wall by a calibration hose inverted to the lining under water pressure. The lining is cured by circulating hot water. With air inversion, the lining is inverted under air pressure and cured by introducing steam.

Cured-in-place pipe lining can be manufactured to suit many sewer shapes and can accommodate small deformations and changes in the direction of the sewer. The pipe is available in various diameter sizes. Sewers from 100 to 2 400 mm (4 to 96 in.) in diameter have successfully been lined using CIPP, with larger sewers requiring special installation procedures. However, it is more difficult to achieve a consistent cure in larger sewers because of the wall thickness and the volumes of water required.

Depending on size, up to several thousand feet of CIPP can be installed in a single insertion. Full bypass pumping or diversion of flows is necessary, although insertion is typically completed within 24 hours.

The lining usually will be designed to fit closely to the existing sewer, making annular grouting unnecessary. In brick sewers where there is missing mortar, excess resin will partially repoint the brickwork. This has the added advantage of providing some bond between the lining and the existing sewer, although this should not be relied on to enhance structural performance.

There are circumstances where secondary grouting is required, as in consolidating the brickwork in multi-ring structures or grouting areas of loosely compacted material or voids adjacent to the sewer pipe. Grouting should be done at specified pressures and the lining designed accordingly.

Preparing the sewer includes removing roots, sediment, and encrustation from the sewer and cutting out intruding connections.

Depending on the primary function of the lining (structural or infiltration reduction), there are certain limitations on use. Deformation of the existing sewer should be limited to 10%. Linings can only be used in sewers that have concave walls with no flat sections or corners. There should be no significant changes of cross section between adjacent manholes. Groundwater intrusion may be transferred to unlined side sewers. Also, it may be difficult to secure competitive bids for the construction contract.

Standard specification for CIPP linings are provided by the American Society for Testing and Materials (ASTM 1216). The first issue of this standard is based on the Insituform® process, and modifications are necessary when other manufacturers' systems are used. Because there are no standard material specifications in the U.S. for these linings, one must rely solely on the guidelines of the manufacturers and the engineer's experience.

Cured-in-place pipe is flexible and requires support from the surrounding material to ensure its long-term stability. Lining with CIPP should be designed using flexible design theory.

NU-PIPE AND U-LINER DEFORMED PIPE. These deformed pipe systems involve inserting HDPE or PVC thermoplastic U-shaped pipe that is expanded in the sewer to form a tight fit with the existing sewer (Steketee,

1988, and Ledoux and Catha, 1988). The installation procedure is illustrated in Figure 7.5.

The pipe is either extruded in a folded form or extruded in the conventional circular form and subsequently folded using a thermomechanical tool. The result is a pipe that can fit tightly inside the sewer but is easily installed because its cross section during installation is significantly less than the final section. The pipe is expanded by processing with heat and pressure in the form of steam to reround the lining.

The lining is available in diameters ranging from 50 to 450 mm (2 to 18 in.) for HDPE and 100 to 300 mm (4 to 12 in.) for PVC. Depending on the diameter, the lining can be installed in continuous lengths of up to 1 500 m (5 000 ft).

Lining can be structurally designed as a flexible pipe with the existing sewer and surrounding ground providing side support. There is some concern regarding the performance of linings in sewers with deflected joints because it is not known how the material will react to the eccentric loadings.

ROLLDOWN AND SWAGEDOWN DEFORMED PIPE. Lining systems using these deformed pipes reduce the diameter of conventionally formed circular PE pipe to allow installation inside an existing pipe. Rolldown and swagelining are shown in Figures 7.6 and 7.7, respectively. When the diame-

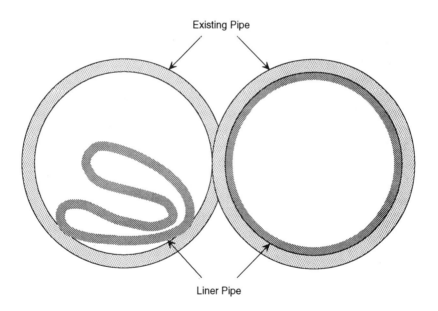

Figure 7.5 Deformed pipe.

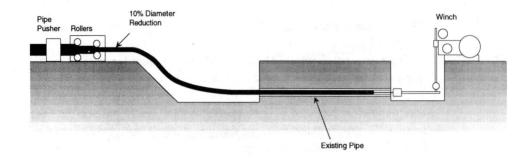

Figure 7.6 General arrangement of rolldown and insertion.

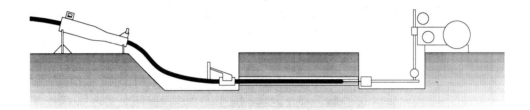

Figure 7.7 Swagelining.

ter is returned to its original size, it forms a tight fit with the existing sewer, increasing the hydraulic capacity of the pipe.

These systems originally were conceived for gas main rehabilitation, and are variants on the conventional continuous sliplining technique.

In the rolldown system, the pipe is cold rolled on site to reduce its diameter sufficiently to allow conventional sliplining insertion (Horne *et al.*, 1987). After insertion, the lining is pressurized to restore the pipe to its original size, resulting in a tight fit inside the pipe. The rolldown machine has five sets of rollers set to progressively reduce the diameter as the pipe is pulled or pushed through.

With swagelining, the fusion-welded PE is pulled through the swagelining machine where it is heated and then passed through the swaging die. This reduces the pipe's diameter by 7 to 15%. The new pipe then is pulled through the old pipe in the conventional sliplining manner. Once in place, the new pipe expands to form a tight fit. The pipe can be pressurized to speed up the process. These pipes are available in diameters ranging from 50 to 600 mm (2 to 24 in.). The lining can be structurally designed as a flexible pipe with the existing sewer and surrounding ground providing side support.

SPIRAL-WOUND PIPE. Using spiral-wound pipe involves inserting a pipe fabricated at the bottom of a manhole or access shaft (Menzel, 1988). A PVC strip is pulled through a winding machine that incorporates a series of rollers that form the pipe to the correct size. The spiral joint is made using either an interlocking PVC clip, twin rubber gaskets, or a water-activated polyurethane adhesive. The pipe is literally wound into the existing pipe, and the annular space is grouted with a cement-based grout. Three basic forms are available: remote spiral winding for non-person-accessible sewers, person-entry spiral winding for larger sewers, and manually installed panels.

With spiral-wound pipe, there are minimal surface disruptions and infinitely variable pipe sizes. However, the continuous pipe joint provides a potential route for infiltration and root penetration. If significant infiltration is present on large-diameter installations, control measures may be necessary before and during installation.

The lining can be structurally designed as a flexible pipe with the existing sewer and surrounding ground providing side support.

Segmental Linings. Segmental linings are suitable for sewers larger than 1 050 mm (42 in.) in diameter and are available in a variety of shapes. For structural reasons, joints should not be located at or near the crown. Segments may be installed through existing manhole chambers with the cover slab removed or, where such access is inadequate, through specially constructed access shafts.

Because it is necessary to install the segments in a dry or nearly dry sewer, bypass pumping or flow diversion usually is required. The annular space between the lining and existing sewer is filled with a high-strength grout installed at relatively low pressures (34.5 kPa [5 psi]). Segmented liners are made from an assortment of materials with varying resistance to corrosion. Thinner linings can be incorporated with cast-in-place concrete to provide corrosion protection, and pressure grouting of the surrounding ground can be applied after installation.

All segmental linings are designed as composite linings, acting together with the existing sewer and grout.

FIBER-GLASS-REINFORCED CEMENT. These linings are prefabricated thin panels designed for sewers 1 050 mm (42 in.) or more in diameter. After the existing pipeline is cleaned and dewatered, the segments are provided in 1.2- to 2.4-m (4- to 8-ft) lengths that overlap at the circumferential and longitudinal joints. Segment ends may be predrilled for fitting to the pipeline by screws or impact nail gun. Segmented rings are anchored on spacers and, after final assembly, the annular space is cement grouted. It may be necessary to provide internal support during the grouting operation.

Fiber-glass-reinforced cement (FRC) linings provide flexibility to accommodate variations in grade, slope, cross section, and deterioration. The linings can be designed to support ground and traffic loads; the bond is the limiting factor. The smooth interior surface can improve hydraulic capacity of a brick sewer but, as with other sewer materials, over extended periods the surface is prone to sliming. The segmented sections are lightweight and easy to handle but installation is labor intensive and slow.

FIBER-GLASS-REINFORCED PLASTIC. Though similar to FRC liners, these linings are more corrosion resistant and less permeable to wastewater and gases. The segments also are lighter and easier to install.

REINFORCED PLASTIC MORTAR. Reinforced plastic mortar (or RTRP) is similar to FRP liners. Though using the same resin and fiber glass designs as FRP lining, RPM uses a fine sand and resin mortar in the matrix and uses continuous rovings.

POLYETHYLENE. Polyethylene sheets are placed in the pipeline covering the upper 270- or 300-deg inner circumference. The sheets normally are 6 m (20 ft) long and placed in one or two segments circumferentially. The sheets are affixed to thoroughly cleaned pipe, using stainless steel molly bolts or pins connected to the remaining wall section. Circumferential and longitudinal jointing between the sheets is done with a special fusion-welding system. Extrusion or butt welding is used for sealing the ends, and this method also can be used for protecting manholes and other structures. This technique is not considered structural and is used primarily for corrosion protection. The system is not well suited for controlling infiltration.

POLYVINYL CHLORIDE. Polyvinyl chloride sheets are placed in the pipeline by methods similar to those for PE. The method is nonstructural and used primarily for corrosion protection.

STEEL. Welded steel pipe in 6-m (20-ft) lengths with plain ends and cut longitudinally are lowered into the access pit. The steel wall is folded over itself, reducing the overall diameter and permitting easier entry. A mobile carrier mounted on wheels transports lining sections down the pipeline, where they are expanded and positioned concentrically with the existing pipe, using bars welded to the outside of the lining or eccentrically on the bottom of the existing pipeline. The longitudinal seam is closed by butt welding, and the ends of adjacent segments are either lap or butt welded in place. After positioning the lining segments, the space between the old and the new pipes is grouted, typically through grout holes in the lining.

Structural and Nonstructural Coatings. *In situ* coatings may be used to extend the life of an existing sewer by increasing its strength or protecting the existing fabric from corrosion or abrasion. Coatings also may be used to improve hydraulic performance.

Reinforced shotcrete and cast-in-place concrete are proven coatings, though only suitable for sewers larger than 1 200 mm (48 in.) in diameter. *In situ* coatings are difficult to apply if significant infiltration is present, and control measures may be needed before and during installation.

REINFORCED SHOTCRETE. Shotcrete coating is the application of concrete or mortar conveyed through a hose and pneumatically projected at high velocity onto a surface. Shotcrete includes both wet- and dry-mix processes, but the term *shotcrete* usually refers to the wet process; the dry-mix process typically is referred to as gunite. Shotcrete and gunite linings can provide structural strength and improved hydraulic performance with some improvement in corrosion resistance.

In situ shotcrete and gunite linings are used in large-diameter (1 200-mm [48-in.] and larger) sewers and manholes and other structures. Various segmental linings can be incorporated to the finished cement to provide corrosion protection. A typical section through a gunite or shotcrete lining is shown in Figure 7.8.

Shotcrete and gunite processes use similar materials and incorporate steel or mesh to limit cracking and provide structural strength. Various latex polymers can improve bond strength, reduce absorption and permeability, and increase chemical resistance. Typical mixes are shown in Table 7.7.

Shotcrete needing no more than 11 to 13 L (3 to 3.5 gal) of water per sack of cement will have low shrinkage and up to twice the in-place density or strength of cast or hand-placed structural concrete. This high density and low permeability reduces the penetration of water to less than 5%. Shotcrete usually is more chemically resistant to acids than ordinary concrete, and the percentage of voids can be less than 50% of cast-in-place concrete.

CAST-IN-PLACE CONCRETE. Lining with reinforced or nonreinforced concrete is an effective rehabilitation method for a variety of conduit shapes. Slip- or fixed-form construction practices are used for concrete placement. This method is used in large-diameter (1 200 mm [48-in.] and larger) sewers with adequate access for materials to be handled effectively. The sewer must be thoroughly cleaned and dewatered before rehabilitation.

When used, the designed steel reinforcement is affixed to the existing pipe. The forms are positioned to provide the finished wall section before the concrete is placed. Structurally reinforced or nonreinforced concrete can be designed for rehabilitation of an existing pipeline, the structural condition of which determines if steel reinforcing is required. Reinforcing steel can be single or multiple layers of preformed, welded wire mesh or hand-placed cages attached to the existing pipe wall by threaded inserts. The new concrete wall can vary in thickness, depending on the design. The concrete mix also can

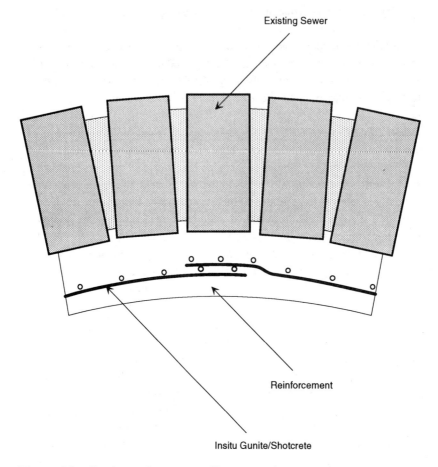

Figure 7.8 Gunite or shotcrete wall construction.

vary and can include corrosion-resistant additives and cements. Segmental liners can be incorporated to the form for total corrosion protection, and precast invert panels often are needed.

Polyvinyl chloride or other plastic liners may help provide corrosion protection in aggressive environments, such as those with high concentrations of hydrogen sulfide.

*P*IPELINE REPLACEMENT OR RENEWAL

It is not always possible or desirable to rehabilitate an existing pipeline, and hydraulic capacity frequently is a problem. While renovated systems can exhibit lower friction coefficients and small increases in capacity, large pipes

Table 7.7 Typical shotcrete and gunite mixes.

Wet mix (shotcrete) design	
Material	**Quantity/cu yd**
Cement (Type I, II, or V)	750 lb[a]
3/8-in.[b] Pea gravel	800 lb
Concrete sand	2 040 lb
Super plasticizer (optional)	Standard dosage
A.E.A.	As required for 4–7% air
Liquid accelerator (optional)	3–5% weight of cement
Water	286 lb
Dry mix (gunite) design	
1 part cement (Type I, II, or V)	
3 to 4.5 parts concrete sand	
Accelerator/admixture (optional)	
Maximum 4 gal[c] water	

[a] lb × 0.453 6 = kg.
[b] in. × 25.40 = mm.
[c] gal × 3.785 = L.

frequently are required. Similarly, if the existing pipe network is not appropriately located to serve both existing and proposed infrastructure, it may be necessary to replace or supplement the existing pipes.

Traditionally, open cut and tunneling methods have been used in such situations. However, these techniques have many limitations. Trenching techniques are disruptive, particularly in urban areas, while tunneling is economical only at greater depths and in larger sizes.

TRENCHLESS REPLACEMENT. In recent years, trenchless replacement and construction techniques have been used more frequently, including pipe bursting, microtunneling, directional drilling, fluid jet cutting, impact moling, impact ramming, and auger boring. Minimum excavation techniques, such as narrow trenching, are also being used more in less-congested environments.

None of these systems are appropriate in all circumstances, and limitations include the proximity of other buried utilities, the presence of boulders in the soil, and solid rock. As the use of nontraditional techniques becomes more common, such difficulties should be overcome.

Pipe Bursting. Pipe bursting is a method for inserting a new pipe of equal or larger diameter to an existing pipeline by fragmenting the existing pipe work and forcing the material into the surrounding soil. The new pipe then is inserted into the enlarged hole.

The concept was developed in the U.K. to replace cast-iron, gas-distribution pipes, and the system is becoming more widely accepted (Reed, 1987 and 1988, and Decker and Larson, 1988). Bursting the pipe is accomplished by using pneumatic or hydraulic bursters. Depending on the system, the burster is either directionally guided or towed by a winch, the new pipe being either towed or jacked immediately behind the burster. The system is illustrated in Figure 7.9.

Several designs of bursting head have been developed, consisting in their simplest form of a series of tubes increasing in size from the front to the back of the machine. Each size increase is accomplished by a conical-shaped transition piece, and special expanding knuckles are fitted to the burster to crack the joints. Hydraulic, expanding bursting machines also have been developed.

A sleeve pipe is towed directly behind the burster to protect the final pipe from scouring damage by broken pipe fragments. After the bursting operation is completed, the sleeve is lined with HDPE, which forms the final pipeline. Other final pipe materials also have been used with varying success, including vitrified clay, FRP, and concrete pipes.

Pipe bursting is suitable for replacing pipes made of brittle material such as vitrified clay, unreinforced concrete, asbestos cement, some PVC, and cast iron. It is not appropriate for the replacement of steel, ductile iron, reinforced concrete pipes, PE pipes, or composite. The sizes needed for this system are shown in Table 7.8, and the designer should verify them before selecting a system.

The maximum length between insertion pits will depend on factors such as the existing pipe material, joint design, pipe surround, and ground conditions. Typically, the length of insertion should be limited to 60 to 90 m (200 to 300 ft).

Pipe surround has a noticeable effect. The effect of concrete surround is dependent on the extent of the concrete, its strength, and the bursting capability of the bursting equipment. The material surrounding the pipe has to accommodate the necessary enlargement of the pipe without inducing frictional

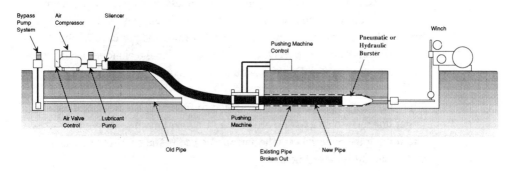

Figure 7.9 Pipe bursting—typical site layout.

Table 7.8 Pipe bursting replacement method.

Internal diameter of existing pipe, in.[a]	Internal diameter of size-for-size replacement pipe, in.	Internal diameter of oversized replacement pipe, in.
9	9	11
		12.4
		13.8
12	12	12.4
		13.8
		15.7
15	—	15.7
		15.7
18	17.7	—

[a] in. × 25.40 = mm.

effects that cannot be overcome. The large roots of trees in the vicinity of a sewer can reduce the rate of progress.

The ground condition surrounding an existing sewer has great influence on the success of pipe bursting, with the ease of replacement typically increasing as the soil type changes from sands and gravels to clays. Bursting in sands or cohesionless materials (particularly wet sands) tends to result in shorter drive lengths than in clays because of the high friction effects caused by almost immediate soil relaxation into the replacement pipe.

ADVANTAGES. Principal advantages of pipe bursting include

- Speed of installation, with rates of up to 37 m (120 ft) of installed pipe per week, including site setup and pipe welding. Insertion rates often are quoted at mm/s (ft/min), which does not account for setup time, including pipe welding (if continuously welded pipe is used), excavation of the insertion pit and laterals, reconnection and ancillary works. Rates of between 2.5 and 20 mm/s (0.5 and 4 ft/min) have been achieved, although hard ground conditions can reduce the rate.
- Reduced restoration costs, particularly in dense urban areas.
- Improved environmental and traffic management. However, experience has shown that this may be overstated. Consideration should be given to the location of lead-in trenches, the number of open holes necessary for the reconnection of laterals, and the space required for laying out the pipe.

- Cost saving with respect to disruption of utility services. This is somewhat offset by concern about the effects of the vibrations caused by the system (see below).
- Unit replacement costs typically 20 to 40% lower than conventional open-cut methods.

DISADVANTAGES. Pipe bursting equipment has a limited size capability, although as the system develops and becomes more widely available, the size range should increase. The system is only suitable for brittle materials and cannot always cope with concrete-encased pipes. This can be a problem if the pipeline construction is not known or only isolated sections are encased in concrete.

In sewers constructed near bedrock or on piles, the forces exerted by the bursting machine will be uneven and the equipment will tend to rise. Large boulders located near the existing pipeline may cause similar problems.

Rates of progress are dependent on the pipe material and ground conditions, making it difficult to predict construction times.

To limit damage, all lateral connections should be excavated and disconnected before bursting and reconnected only after the new pipe is fully installed. This may be a problem where large numbers of laterals are involved, and consumer service may be cut for an extended period if construction problems occur.

One of the greatest disadvantages is the risk of damage to adjacent services, structures, and the ground surface, particularly where surfaces are paved. The expansive action of the burster results in movements in the material surrounding the pipeline, the amount of movement depending on the soil material, the degree of expansion, and the type of pipe encasement. Movement is transmitted through the surrounding material, inducing strains and bending forces in transverse crossings and other adjacent pipelines. These forces can cause pipeline movement and, depending on the magnitude of the force and the condition of the pipe, could cause structural damage or leakage at a joint.

A significant amount of research has been completed in this area, leading to pipe proximity charts (Leech and Reed, 1989, and Briassoulis *et al.*, 1989). Shallow operations will be influenced significantly by the proximity of the ground surface, leading to a generally upward movement during expansion, with local and intense disturbance above the bursting process. In deeper expansions, as is typical for sewers, disturbances are more likely to be contained by compression in the surrounding soil. Expansions in homogeneous soils will be radial, while ground movement in the hard layer below the expansion (bedrock, piles, trench bottom with weaker backfill) will be directed upward. The degree to which movement is localized and contained by volume change will depend on the strength and compressibility of the backfill.

Surface disturbance will vary from slight heave with little visible damage to pronounced heave with severe surface cracking and opening of existing joints and defects. Much of the disturbance produced during bursting will be transient, with some subsequent subsidence.

The most appropriate method of predetermining heave is to compare proposed work with similar completed contracts.

Contractors should be able to direct the engineer to new developments in this constantly changing field.

Microtunneling. Microtunneling was pioneered in Japan, where the prospect of intrusive excavation in heavily populated urban areas to provide first-time sewerage facilities led to restrictions on open-cut construction (Water Research Centre, 1988, and Thomson, 1991). While tunneling methods were the obvious choices for the larger mains, there were no equivalent techniques for pipes with 150- to 900-mm (6- to 36-in.) diameters. Assisted by the soft or granular nature of the soils, remote-controlled, steerable machines were developed for trenchless construction of these smaller sizes.

In the early 1980s, equipment was developed to address the problems of first-time sewerage in urban areas of Germany. The stimulus came from the need to deal with high groundwater levels with minimum effect on adjacent structures, combined with a strong environmental lobby.

Microtunneling has a high unit-construction cost compared to traditional, open-trench systems. Unless the cost of disruption to utility services is high, microtunneling will not be competitive.

In the U.S., increased workloads have increased awareness in the industry of the benefits of microtunneling, and the technology should become more competitive as it develops to deal with more varied ground conditions.

When dealing with I/I, microtunneling is an option in urban areas where an existing pipeline is to be abandoned or requires upsizing but is still below the size for conventional tunneling. Microtunneling is discussed in this manual as a possible technique for inclusion in a complete I/I removal strategy.

Microtunneling is a method of excavation to permit the installation of a pipe by pipe jacking, which requires a remote-controlled, steerable tunnel-boring machine in 150- to 900-mm (6- to 36-in.) diameter range. The machines fall into two categories, according to their method of removing the spoil from the tunnel face.

AUGER SYSTEMS. Auger systems incorporate a series of helical augers within the tunnel boring to transport soil from the cutting head to the jacking pit or starting shaft, where it is removed (Figure 7.10). The cutting head is connected directly to the auger flight and is driven from an electric motor located in the jacking shaft. A choice of cutting heads is available for differing soil conditions. Crushing heads also are available to

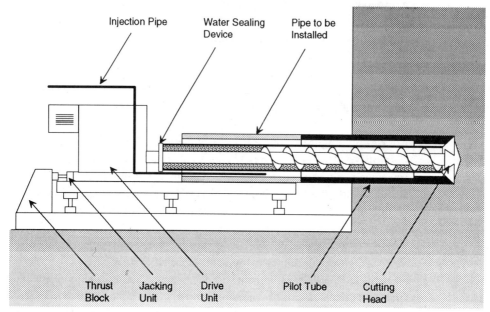

Figure 7.10 Microtunneling (auger system).

deal with cobbles, and recent developments include equipment to remove existing sewers.

By activating several small hydraulic steering jacks near the front of the machine, directional control is achieved using a laser beam emitter located in the jacking pit and a target located on the rear of the tunneling machine.

Groundwater at the tunneling machine level can result in overexcavation. This can be limited by the injection of slurry, water, or compressed air at the cutting head.

SLURRY SYSTEMS. A small-diameter discharge pipe installed within the lining of the tunnel carries the soil removed by the slurry shield machine directly to a treatment plant at ground level (Figure 7.11). The slurry liquid normally consists of a bentonite and water mixture, although water alone may be suitable in some soils or machines. In soils containing cobbles or stones, a crushing head is needed to grind the material down to a consistency suitable for passing through the small-diameter discharge pipe.

The drive to the cutting head is applied directly from a motor and gearbox located at the front of the tunneling shield. Steering and control takes place in a manner similar to that used in the auger system.

PIPE INSTALLATION. The pipeline itself is jacked behind the microtunneling machine, to provide the forward motion of the machine. Additional lengths of pipe are added at the insertion pit, the cycle continuing as the com-

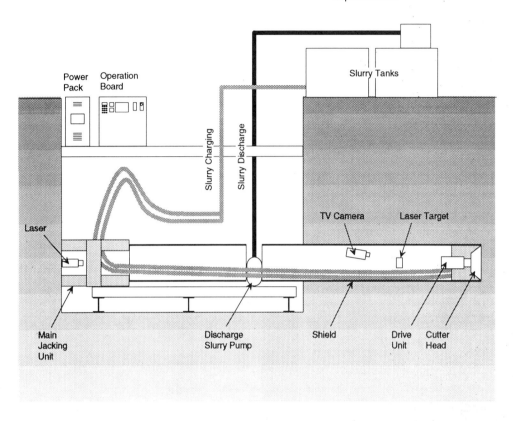

Figure 7.11 Microtunneling (slurry system).

plete pipe string is pushed or jacked forward. Direction is controlled by the microtunneling machine as described above. Pipes are manufactured from a variety of materials, including concrete, vitrified clay, fiber-reinforced plastic, fiber-reinforced cement, asbestos cement, ductile iron, and steel.

In most cases, special joints have been developed to provide a smooth internal and external profile, the joints being contained within the pipe wall. The joint also should be designed to cope with the jacking forces, and be capable of small deflections without leakage. A jacking pipe also should be able to withstand direct and eccentric jacking forces, soil and traffic loads should be supported safely, and the pipe material should have sufficient durability for the sewer environment.

ADVANTAGES. Among its advantages, microtunneling allows for close control of the line and grade, and is capable of installing a permanent lining in standard lengths between access points. Also, it is able to function in difficult ground conditions and in congested areas with minimal surface disruption at depths greater than 4.5 to 6 m (15 to 20 ft).

Pipeline Rehabilitation Methods

DISADVANTAGES. Disadvantages of microtunneling include high initial capital cost, requiring extensive support services. Larger obstacles such as boulders can cause considerable difficulty. The system is only suitable for I/I applications where deep pipes require replacement with larger pipes, or where there is insufficient scope for reducing I/I in upstream sewers, creating a need for additional interceptor sewers in the downstream reaches.

OTHER TRENCHLESS SYSTEMS. Other trenchless systems, while significant in their development, do not directly apply for I/I problems. Such techniques include steerable systems, such as directional drilling and fluid jet cutting, and nonsteerable systems, such as impact moling in virgin ground, impact ramming, and auger boring.

Directional Drilling. This technique is used predominantly for the installation of long, vertically curved pipelines, usually under bodies of water such as rivers, estuaries, and canals (Reynolds and Szczupak, 1987). Using substantial surface equipment and being capable of drives to more than 1 000 m (3 300 ft), the technique is best suited to major schemes that need expensive and heavy equipment (Figure 7.12).

In this technique, a small-diameter pilot hole is drilled in a shallow arc. A washover pipe sightly larger than the pilot tube follows the drill string, acting both as temporary support and a method of reducing friction on the drill string before enlargement. The completed pilot bore is enlarged using back-reaming techniques until large enough to receive the final pipe, which is normally steel, although polyethylene and bundles of pipes also have been used.

Fluid Jet Cutting. This technique uses a remote-controlled and guided tool that incorporates high-pressure (6 900 to 27 600 kPa [1 000 to 4 000 psi]) slurry jet nozzles for cutting soil to provide a pilot bore (Kirby, 1991). The final tunnel, available in sizes between 50 and 350 mm (2 and 14 in.) in diameter, can be up to 120 m (400 ft) long and 10 m (34 ft) deep (Figure 7.13). The process is monitored from the surface, and drilling direction adjustments are made in both the vertical and horizontal planes. Accuracy is within plus or minus 150 mm (6 in.), making the system suitable for pressure lines and gravity lines, where the tolerance to line and level is not critical.

Impact Moling. Moling uses compressed air to drive a cylindrical percussive hammer through the soil to form a hole. The soil is simply consolidated from the pipeline, and not removed. The system is nonsteerable and thus not appropriate for gravity lines.

Impact Ramming. This is a development of impact moling where a large impact mole in a drive pit is used to hammer a steel casing into the ground. Soil enters the open-ended sleeve and is removed by jetting, mechanical cutting,

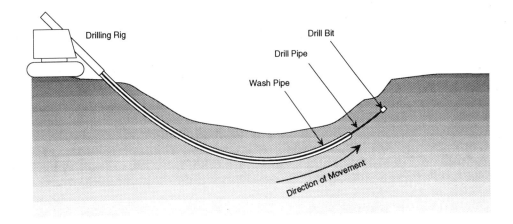

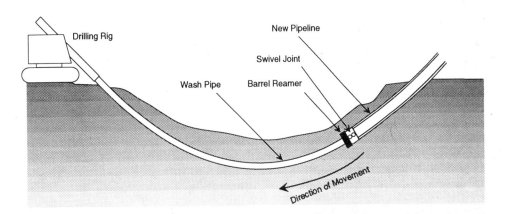

Figure 7.12 Directional drilling.

or simply pushing out the soil plug. The system is limited in drive length to approximately 30 m (100 ft), and is only applicable to medium and larger crossing-type work over modest lengths.

Auger Boring. This technique consists of a rotary cutting head followed by an auger flight for soil removal. The equipment has a wide use in crossing-type work, but is not generally appropriate for I/I problems.

CONVENTIONAL REPLACEMENT. The ultimate solution to I/I problems is complete replacement of the pipeline. To some degree, this is achieved with some of the lining and trenchless replacement options in that a

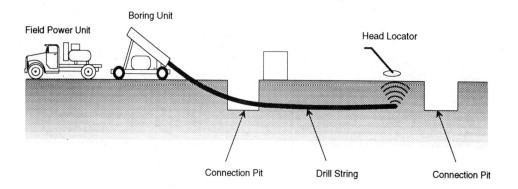

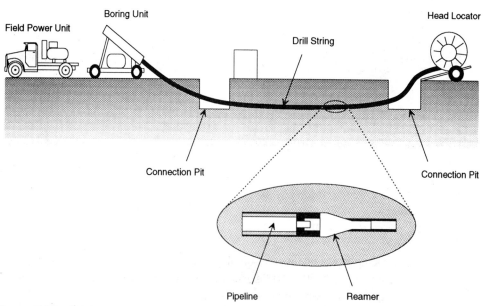

Figure 7.13 Schematic arrangement of fluid jet cutting system.

new pipe is provided, but the inflow part of the equation is not solved. Complete replacement allows the option to completely divert and remove extraneous flows.

Traditional, open-cut techniques are often economical in relatively open suburban and rural environments, where the disruption easily can be accommodated. However, in heavily congested urban areas, or where sewers are deep, lining and trenchless systems become more competitive. Tunneling only becomes an option when large sewers with severe infiltration problems are to be rehabilitated.

*M*AINTENANCE AND REPAIR

Maintenance and repair can be considered part of pipeline rehabilitation and encompass a number of techniques.

CLEANING. The purpose of sewer cleaning is to remove foreign material from the sewer and generally is undertaken to alleviate one of the following conditions (Knott, 1989 and 1990):

- Blockages (semisolid obstructions resulting in a virtual cessation of flow). These generally are dealt with on an emergency basis, although the underlying cause can be treated preemptively.
- Hydraulic capacity. In some cases, sediment, roots, intrusions (connections or other foreign bodies), grease, encrustation and other foreign material restrict the capacity of a sewer, causing surcharge or flooding. Cleaning the sewer may alleviate these problems permanently, or at least temporarily.
- Pollution caused by either the premature operation of combined wastewater overflows because of downstream restrictions to hydraulic capacity or pollution caused by the washing through and discharge of debris from overflows during storms.
- Odor caused by the retention of solids in the system for long periods resulting in, among other things, wastewater turning septic and producing hydrogen sulfide.
- Sewer inspections, where the sewer needs to be cleaned before inspection. This requirement most often occurs when using in-sewer CCTV inspection techniques.
- Sewer rehabilitation where it is necessary to clean the sewers immediately before the sewer being rehabilitated.

The degree of cleaning required is contingent on the purpose of the cleaning and the extent of rehabilitation planned. There are times when CCTV work may be done in advance of cleaning. When internal grouting is intended, the pipe joints must be cleaned for satisfactory seating of the packer equipment and access to the joint. The success of other types of rehabilitation also depends on the cleanliness of the pipelines.

Conditions such as broken pipe and major blockages prevent cleaning from being accomplished, and cleaning should not be attempted or continued if the pipeline has serious corrosion or structural problems.

Common cleaning methods include jet rodding, manual rodding, winching or dragging, cutting, and manual or mechanical digging. The method usually is determined in advance and is normally contingent on the pipe type and size and on the conditions expected in the pipe.

Jet Rodding. This method depends on the ability of high-velocity jets of water to dislodge materials from the pipe walls and transport them down the sewer. Water under high pressure (approximately 14 000 kPa [2 000 psi]) is fed through a hose to a nozzle containing a rosette of jets sited so the majority of flow is ejected in the opposite direction of the flow in the hose. These jets propel the hose through the sewer and dislodge the materials on the sewer walls. A range of nozzles is available to cope with the different pipe diameters and materials encountered. The hoses, nozzles, water supply and necessary pumps usually are incorporated in a purpose-built vehicle. Equipment for removing and storing discharged material also is provided on some cleaning units.

Rodding. This method is generally a manual push–pull technique used to clear blockages in smaller-diameter, shallow sewer systems typically not exceeding 250 mm (10 in.) in diameter or 1.8 m (6 ft) in depth. For sewers greater than 250 mm (10 in.) in diameter, the rods tend to wander and are not very effective. The distance from the access point is limited to approximately 18 m (60 ft).

Winching or Dragging. This is a technique where custom buckets are dragged through the sewer and the material deposited into skips. The winch can be powered or hand operated, and the system used in sewers up to 915 mm (6 in.) in diameter and 50% silted.

Cutting. This method generally is used for removing roots from sewers. High-pressure water jet cutters have been developed in Europe for removing more solid intrusions, such as intruding connections. Some of these have been fairly successful, but care is required to eliminate damage to the existing sewer structure. This equipment is likely to become more commonplace.

Manual or Mechanical Digging. Traditionally used in larger-diameter sewers, this method involves manually excavating the material and placing it in buckets for removal. As the sewer system can be hazardous, the technique now is used infrequently. High-pressure jet equipment also can be used manually in larger sewers.

ROOT CONTROL AND REMOVAL. Root intrusion increases the rate of infiltration to a sewer by expanding an existing opening. This can allow greater quantities of the surrounding soil to enter through the defect, further weakening the structure and ultimately leading to breakage and collapse of the sewer structure. The roots themselves also can have a detrimental effect on hydraulic conditions in the sewer by creating a local flow restriction. This increases the possibility of surcharge by screening out the solids, further restricting flow and reducing velocity. The frequency of surcharge also has an effect on the structure's rate of deterioration.

Control and removal of roots in a sewer is an important and ongoing maintenance operation, and an effective root control program requires an understanding of root growth.

Roots grow toward moisture and nutrients by a continuous process that adds cells, one at a time, to the tip of the root. This cell-by-cell growth enables roots to penetrate small openings in sewer pipes. Because roots seek moisture and nutrients, root intrusion tends to be a problem only in sewers installed in soils where moisture is limited or has seasonal variations, including pipelines continuously or seasonally above the groundwater table. Root intrusion is more common in service connections, which are shallower and frequently not as well constructed as sewer mains.

The most effective root control method is to prevent roots from entering a sewer in the first place. Installing watertight lines that are free from imperfections and will not crack, break, or deteriorate during service is the ideal solution. This may require that materials and construction methods meet or exceed current standards, and also may necessitate increasing on-site inspection during the installation of pipelines and service connections.

The potential for root intrusion also can be reduced by discouraging tree planting near sewer lines.

When roots have penetrated the pipeline, the general course of action involves mechanical removal or chemical treatment. Typically, maintenance crews cut the roots using a sewer rod and an auger tool. Although this corrects the immediate problem, in the long term, roots frequently grow back more vigorously after each cutting operation. Root cutting should be followed by chemical treatment or by flooding the pipeline with scalding water to retard root regrowth.

Herbicides for chemical treatment generally are applied by flood treatment (soaking), spraying, or foaming. Flood treatment permits control of chemical concentration and contact time during application, ensures that the herbicide penetrates the roots, and may kill some roots outside the pipe. The major disadvantage is that this method interrupts sewer service for approximately one hour. Using standard hydraulic cleaning equipment to spray herbicides on the roots results in a short contact time and has not yielded significant long-term root control.

The use of foam herbicides is increasing because they are relatively easy to apply and effective in treating root intrusion in building services without the problems generally associated with the flood treatment technique.

Chemical grouts used to seal joints may also contain herbicides to control root intrusion. An advantage of combining grouting materials with herbicides is that it addresses two problems, infiltration and root intrusion, using the same rehabilitation technique. If the pipeline already is subject to moderate or heavy root intrusion, however, the roots must be removed before the chemical grout is applied.

In extreme cases, where the sewer structure is also defective, the removal of roots may be accompanied by relining or replacement of the sewer line to prevent further root growth.

POINTING. Conventional repointing techniques may be used in brick or masonry sewers to reduce infiltration or replace deteriorated mortar. The technique frequently is associated with internal or external grouting designed to restore side support to the structure. The existing sewer still must be physically intact, with no deformation, disturbed bricks, or closed joints. The mortar can be applied manually and troweled into place or delivered to the point of application using pressure pointing equipment. The equipment can be operated on the surface or within the sewer, and delivers premixed mortar to the point of application. The term "pressure" applies to the method of delivery and not the actual application of the mortar. Pressure pointing is a faster process than hand pointing and is more often used when longer lengths of sewer need attention.

Full bypass pumping or flow diversion is required when working below the normal flow line with a need to control infiltration to avoid detrimental effects on the bond between the mortar and brickwork. Sewer walls should be water jetted or manually cleaned to remove encrustation, sediment and slime, and mortar should be removed to a depth of approximately 1.5 times the joint width.

Traditional sand and cement mortars and special pressure pointing mixes typically are used, although the mortar must be of a similar strength to the original as spalling of the brickwork may occur if stronger mortars are used.

INTERNAL GROUTING. Internal, or chemical, grouting is the most common method for sealing leaking joints in structurally sound sewer pipes. Using special techniques and tools, chemical grouts can be applied to pipeline joints, manhole walls, wet wells in pumping stations, and other leaking structures. Small holes and radial cracks also may be sealed by chemical grouting.

Small- and Medium-Sized Pipes. A hollow, cylindrical packer with inflatable collars at either end (Figure 7.14) is inserted to the sewer and positioned across a pipe joint. The rubber collars are inflated and the joint tested under pressure. If the joint fails the test, a chemical grout is pumped through the joint to the surrounding ground until the grout solidifies and an acceptable backpressure is achieved. This technique reduces the potential for migration of groundwater infiltration through other unsealed joints in a sewer segment.

In its uncured state, the low-viscosity grout is pumped through the joints and cracks, curing to form a resilient gel. Curing time can be varied from a few seconds upward, and the amount of grout needed to seal the defect depends on the size of the leak and the receptivity of the ground.

The process is controlled and monitored by CCTV. The television monitor, pumps, air compressors, and chemical feed equipment typically are housed in a van that acts as the operations and control center.

The success of the operation can be assessed with a low-pressure air test of the sealed sewer reach when the curing time has elapsed. Pressures must be limited to avoid rupturing the grout seal.

Large-Diameter Pipe Grouting. For grouting large-diameter pipes, pressure grouting or manual placement of oakum soaked with grout may be used. The pressure grouting may be accomplished using either pipe grouting rings or predrilled injection holes.

Grouting with sealing rings (Figure 7.15) requires a small control panel, chemical and water pumps, and various other accessories, depending on the type of sealing grout being used.

To seal joints using grout sealing rings, a worker must enter the line, manually place the ring over the joint that requires sealing, and then inflate the ring to isolate the joint. The sealing rings work on the same principle as small-diameter pipe packers in that sealing grout is pumped into the small void between the pipe wall and the face of the ring.

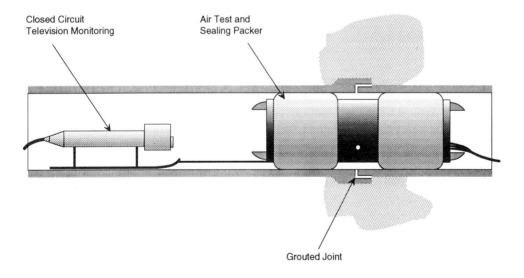

Figure 7.14 Internal grouting equipment.

Pipeline Rehabilitation Methods

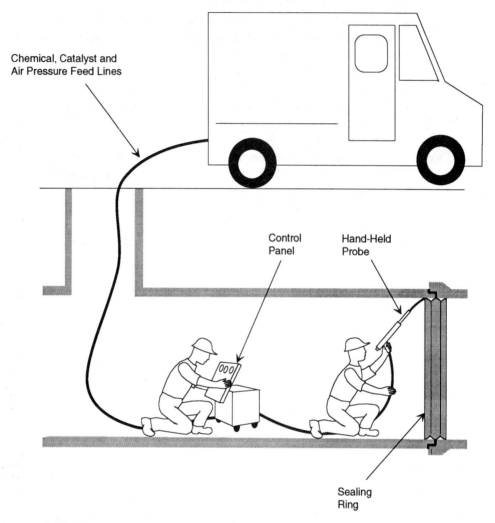

Figure 7.15 Internal grouting of large-diameter pipe joints.

In this method, the grout is pumped through a hand-held probe. As the pressure increases, the grout solution is forced into the joint and surrounding soil. After the catalyst solution is injected, the grout cures, sealing the joint from infiltration.

The probe injection method uses most of the same equipment as the sealing ring method, except that sealing rings are not required. As before, equipment depends on the type of grout being used. The worker enters the line to drill holes, either into or next to the joint. The actual number of holes depends on factors such as size of pipe and length of crack. When the holes have been drilled, various techniques to apply the grout can be used. The sealing grout is injected by special injectors and equipment much like that used with the sealing ring. The injector is connected to this section of pipe through the wall and the chemical sealant is pumped into the joint.

An alternative method is to use a tapered probe at the end of the injector. The probe is manually pushed into the drilled holes and held in place during the pumping sequence. The grout and catalyst are pumped to the probe in separate hoses. Locating the holes near or at the joint enables the grout to enter the holes and fill any voids around the joint.

Materials. Acrylamide gel, acrylate gel, urethane gel, and polyurethane foam are the main types of chemical grouts currently available. Other grouts are being formulated to take their place. The manufacturers should be contacted for current data before a specific material type is specified.

The chemicals necessary to form acrylamide or acrylate gels usually are mixed in two separate tanks and pumped through separate hoses to the pipeline joint to be sealed. One tank is used to mix and dispense a solution of water and the acrylamide or acrylate grouting material. The other tank is used to mix and dispense water and catalyst solution, which initiates the chemical reaction when mixed with the acrylamide solution. Additives can be included in either solution to help control shrinkage, reaction or gel time, and other variables.

From the mixing tanks, the two solutions are pumped through separate hoses to the point to be sealed. The solutions are mixed as they are pressure-injected to the leaking opening, initiating the chemical reaction. This reaction changes the two solutions into a gel almost instantly when the predetermined reaction time has elapsed. The gel time can be controlled from a few seconds to several minutes. The acrylamide and acrylate gels stabilize the soil around the defective joints or cracks by filling the voids.

Urethane grout materials may be used to form either an elastomeric gel, much like the acrylamide and acrylate gels, or a rubberlike foam. Urethane gel is different from the acrylamide or acrylate gels in that water is the catalyst for the urethane gel material. The properties of the reacted gels are also different. Urethane gel seals pipeline joints by forming an elastomeric collar within the pipe joint and by consolidating soils and filling voids outside the joint.

Urethane gel is applied in essentially the same manner as acrylamide and acrylate gels, but with the equipment used for the polyurethane foam grout, which also is activated by water. Different equipment generally is used because it is difficult to remove all traces of the water solutions from the mixing tanks and the chemical feed lines. Any water remaining prematurely triggers the chemical reaction and clogs the equipment or feed lines.

Advantages. Chemical grouting can be less expensive initially than other options for sewer rehabilitation. Sealing a line segment usually takes only a few hours, so with proper scheduling, chemical grouting can be accomplished with minimal traffic interruption. All of the chemical sealant is applied from inside the pipe, eliminating the need for excavation and surface restoration. In addition, damage or interference with other underground utilities such as gas lines, water lines, or telephone cables is minimal.

Limitations. Chemical grouting does not improve the structural strength of a pipeline and should not be considered when the pipe is severely cracked, crushed, corroded, or badly broken. Once applied, some chemical grouts may dehydrate and shrink if the groundwater drops below the pipeline and the moisture content of the surrounding soil is reduced significantly. The amount and duration of the shrinkage depends on factors such as the length of time the groundwater remains below the pipe, the moisture content of the soil while the groundwater level is depressed, the soil type, the type of grout, and the additives included in the grout to control shrinkage. The potential for dehydration of the grout should be considered, as this may reduce its service life.

Some joints and cracks may be difficult to seal with gel grouts when large voids exist outside the pipe joint. Extremely large quantities of grout may be required to seal such a joint, if sealing is possible at all. Other joints cannot be sealed with either gel or foam grouts because they are badly offset or misaligned. Offset joints may prevent the inflatable rubber sleeves of the sealing unit from seating properly against the walls of the pipe, making it impossible to isolate and seal the joint. Chemical grouting of longitudinal cracks in pipe also is not generally feasible because the grout may flow through the crack and leak back to the sewer.

Grout may be washed away if there is significant groundwater movement outside the sewer, reducing the effectiveness of the technique and increasing the amount of chemical needed and, hence, the cost. Grouting typically is not a permanent method of repair compared to lining methods and regular resealing is often required.

Cost and Feasibility. Many factors can influence the cost effectiveness of chemical grouting. When considering this method, the pipeline first should be inspected visually to determine cleaning needs, the extent of root intrusions, and the structural integrity of the pipeline. For grouting to be effective, the pipeline must be relatively free of sand, sediment, and other deposits.

Roots in a pipeline can cause severe structural damage, often making total replacement an attractive option. If joints are sealed but the roots are not effectively removed or inhibited, the root problem is likely to return.

The inside walls of the pipe must be smooth enough to allow proper seating of the packer. If a line has several badly cracked or broken sections, partial replacement may first be necessary.

Potential groundwater migration must be considered when evaluating chemical grouting. The test-and-seal technique limits the opportunity for groundwater that migrates along the pipeline to infiltrate other unsealed joints in a manhole-to-manhole sewer segment. Sealing only the sewer mains, however, may transfer groundwater entry to the service laterals. Because most sewers have been backfilled with granular material, which readily transfers water, sealing only isolated manhole-to-manhole segments in an area may transfer groundwater to other, unsealed manhole-to-manhole segments.

The service life of the grout is an important consideration. Acrylamide grouts have been used successfully since the 1950s to stabilize soils and help control underground water movement in tunnels, dams, dikes, pits, and other underground structures. Polyurethane foam has been used successfully since the early 1970s, and urethane gels since 1980. Grouting is not considered a permanent repair method comparable to linine techniques. Grouting may be required as soon as 5 years from the original application, as opposed to 50-year design lives for lining systems.

Pipe size, joint spacing, and the percentage of joints requiring sealing are factors that determine the cost of chemically sealing a line. The larger the pipe, the higher the cost because of increased manpower, equipment, and material. The number of joints requiring sealing in a manhole-to-manhole sewer segment usually is determined by estimating the percentage of joints that will require sealing, multiplied by the manhole-to-manhole segment length and divided by the known joint spacing. Daily production is lower in larger lines and rerouting of wastewater flow around the sections being sealed often is required. Larger lines are also more difficult to clean in preparation for sealing.

EXTERNAL GROUTING. External grouting rehabilitation methods are performed either from aboveground, by excavating adjacent to the pipe, or from inside the pipe, depending on the pipe diameter. Variations of chemical or cement grouting are appropriate for solving problems of significant groundwater movement, washouts, soil settlement, and soil voids.

Chemical Grouts. Chemical grouts consist of a mixture of three or more water-soluble chemicals that produce stiff gels from properly catalyzed solutions. The grouts produce a solid precipitate, as opposed to cement or cement and clay grouts, which consist of suspensions of solid particles in a fluid. The reaction in the solution may be either chemical or physiochemical, involving the solution constituents either alone or in reactions with other substances encountered. In the latter type, the reaction causes a decrease in fluidity and a tendency for the solution to solidify and form occlusions in channels or fill voids in the material to which the grout has been injected.

Acrylamide base gel is the most widely used chemical grout for curtain walls, especially in fine soils. Acrylamide grout is mixed in proportions that, with proper reactions, produce stiff gels from diluted water solutions.

Before acrylamide grout is applied, five fundamentals should be evaluated: the desired result, nature of the grout zone, application equipment, alternative procedures, and plan of injection.

The grout injection plan should include gel times to be used, quantities of acrylamide, and the probable layout for grout injection points. The work plan can be modified, based on information gained during grouting operations.

There is no known natural soil or rock formation in which a gel will not form, though the injected solution must remain in the grout zone until gelation occurs. In dry soils and flowing groundwater, there is a tendency for the grout to disperse. This can be avoided by saturating the soil before grouting and using short gel times. In soil void work, however, a dry soil mass cannot be stabilized as efficiently as a soil mass below the water table.

Dilution around the outer edges of the grout bulb may occur in wet soils when long gel times are used. The flowing groundwater distorts the normal shape of the stabilized mass and can displace it in the direction of flow.

In turbulent flow conditions, dilution can be reduced by short gel times. In open formations or fissures, solids such as bentonite or cement may be added to the grout solution to help produce a more complete block to flowing water. The most successful application of acrylamide occurs in saturated or partially saturated soils.

The majority of grouting applications are intended to reduce leakage rather than increase strength. For sewer rehabilitation purposes, grouting is effective in reducing or eliminating infiltration but will not significantly improve the structural condition. It can, however, indirectly improve structural integrity by stabilizing the surrounding soil mass.

Cement Grout. This grout consists of a slurry (particulate suspension) of cement and water, with materials such as sand, bentonite, or accelerators added, if necessary. The particulate nature of cement grout normally restricts its use to fractured rock and large-grained soils, where the voids are large enough to facilitate penetration and permeation.

Portland Cement Grout. This grout can be used to form impermeable subsurface barriers but is restricted in application to medium sands or coarser materials because of the larger size of cement particles. Various Portland cement grouts have been used successfully to fill voids and washouts adjacent to sewer pipelines. Type III cement often is selected because of its smaller particle size.

A variety of water-to-cement ratios can be used, depending on subsurface conditions. Strength characteristics generally are not important, as grouting primarily is intended to fill voids surrounding a buried sewer pipeline. Proper design of the water-to-cement ratio results in grout mixtures that are easy to mix and inject and have strengths of 3 450 to 6 900 kPa (500 to 1 000 psi). Clays can be added to cement to form gels and prevent settlement of the cement from suspension. This does, however, result in an ill-defined setting time and a slow strength development. Accordingly, they are not normally used in sewer rehabilitation void grouting when groundwater is present.

Portland cement grouts may also be used as fillers and accelerators in silicate grouts and may be mixed with acrylamide to improve water shutoff capabilities and injection characteristics. For extremely large void-filling applications, other cements, such as pozzolan and fly ash mixes, can be used and are more economical than straight Portland cement or soil and cement mixtures.

Microfine Cement Grout. This type consists of finely ground cement, allowing it to penetrate the fine sands that ordinary Portland cement cannot penetrate. Besides its excellent permeability, it exhibits the required strength and durability, with a set time of 4 to 5 hours. When combined with sodium silicate, with no organics, a 1- to 3-minute gel time can be attained for underground water control.

Compaction Grouting. This type of grouting is the injection of stiff, low-slump, mortar-type grout under relatively high pressure to displace and compact soils in place. Compaction grouting acts as a radial hydraulic jack, physically displacing the soil particles and moving them closer together. The technique is used to strengthen loose, disturbed, or soft soils or for control of settlement. Compaction grouting is used primarily on large pipelines and is applied through the pipe wall to the surrounding soil. Care must be exercised to ensure no damage occurs to either the sewer structure or any adjacent utility equipment.

MECHANICAL SEALING. Mechanical seals consist of a rubber-type seal positioned across a defective joint and held in place by stainless steel retaining bands. The seals are available in a number of widths suitable for joints with differing gap widths. A typical seal is shown in Figure 7.16.

The seals are only suitable for sewers where safe internal access is possible as they are installed manually from inside the pipe. Any grease and scale should be removed from the joint before the seal is positioned. The shape of the seals is designed to reduce their interference with the flow.

POINT (SPOT) REPAIRS. Point repairs may be used to correct isolated or severe problems in a pipeline segment. They are most commonly used to totally correct defects within a line segment, but also can be an initial step in the use of other methods. Most point repairs require excavation and some replacement. Point repairs usually are limited to the replacement of only a short portion of a pipeline or service connection.

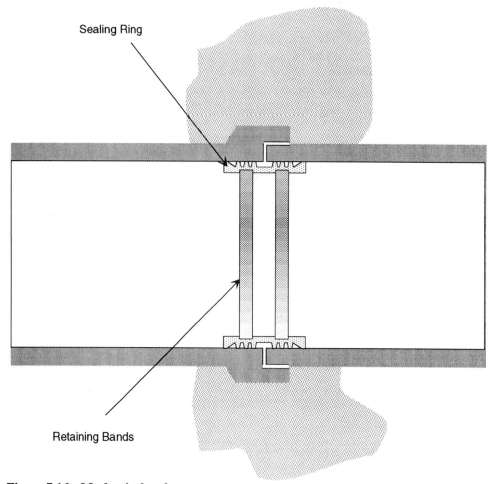

Figure 7.16 Mechanical seal.

Local repairs on brick sewers frequently involve the replacement or repair of longer lengths of sewer than the equivalent pipe sewer because of the need to tie into a competent portion of the structure.

When making point repairs, consideration should be given to the materials and methods used to connect the replacement pipe to the existing pipeline. Flexible couplings often are used to join pipes together. Occasionally, concrete collars and encasement are used.

The repair of pipe connections to manholes is another type of point repair, and is covered in the following section on manhole rehabilitation methods.

*M*ANHOLE REHABILITATION

Manholes are rehabilitated to correct structural deficiencies, to address maintenance concerns, and to eliminate I/I. Manhole rehabilitation may also lessen or prevent corrosion of the internal surface caused by sulfuric acid formed when hydrogen sulfide gas is released from the wastewater to the sewer environment.

Many methods to rehabilitate manholes are currently available (Wade, 1990, and Nelson, 1991), and new products and application technologies continually are being developed. The evaluation of each method should consider the types of problems, physical characteristics of the structure, condition and age, location of the manhole with respect to traffic and accessibility, risk of damage or injury associated with the current condition of the structure, and cost or value in terms of rehabilitation performance.

MANHOLE CONDITIONS. The following sections discuss manhole conditions that could result in manhole rehabilitation.

Structural Degradation. The definition of structural degradation varies with manhole material composition, shape, and size. Structural degradation does not necessarily mean structural failure. For purposes of manhole inspections, structural degradation may be defined as damage to any of the structural components of a manhole. Structural degradation can be caused by movement and displacement or corrosive environments.

MOVEMENT AND DISPLACEMENT. Structural degradation of manholes will occur with three-dimensional displacement and movement. In areas where freeze-and-thaw cycles are common, degradation of the frame seal, chimney, and top portion of the cone can occur. Vertical separation can be dramatic, particularly where manhole frame castings are monolithically encased with rigid and flexible pavement. Horizontal movement of the frame occurs as the encased frame reacts to the thermal expansion and contraction of the surrounding pavement caused by temperature variations.

Three-dimensional movement also can occur to the entire manhole structure from settlement and movement of the ground around the manhole. This differential movement can be pronounced in certain clay or unstable soils. Such loading can impose unbalanced point loadings and increase tensile stress failures. Manholes made of brick and block are particularly susceptible to displacement and joint separation where unstable soils exist. Traffic-induced loads can result in three-dimensional movement of the manhole cover, frame and chimney section, causing cracks and fractures.

CORROSIVE ENVIRONMENTS. When sulfides are present in the wastewater stream because of the natural biodegradation of the wastewater, structural deterioration likely will occur to all concrete surfaces. Factors that control sulfide generation are wastewater velocity, ambient water pH and air temperature within the sewer system, and oxygen availability.

Under extreme conditions, total structural degradation of unprotected precast manholes can occur in fewer than 5 years. Hydrogen sulfide structural degradation of a manhole can be controlled through effective manhole rehabilitation, although wider measures may also be appropriate.

Excessive Infiltration and Inflow. A significant percentage of identified I/I is from defective manholes. Factors that need to be considered when evaluating and quantifying potential I/I from manholes include ground saturation, water table fluctuations, and inspection data on the type and condition of the manhole. A relative ranking of several common manhole I/I sources that contribute to excessive wet weather flows is listed in Table 7.9.

Maintenance. Field conditions that hinder normal maintenance and operations of a collection system should be considered in manhole rehabilitation. These conditions include deteriorated manhole steps, offset frames, buried or inaccessible manholes, other utilities passing through manholes, and nonstructural problems that affect operations and maintenance access to the collection system.

MANHOLE REHABILITATION METHODS. The rehabilitation of manholes can be grouped into the following methods: chemical grouting; coating systems; structural linings; corrosion protection; and manhole frame, cover, and chimney renovation.

Table 7.9 Ranking of manhole infiltration sources.

Source defect	Infiltration and inflow type	Typical range, gpm
Vented cover[a]	Inflow	1.0 – 3.0
Poor cover and rim fit	Inflow	0.1 – 5.0
Frame seal	Inflow	0.5 – 5.0
Cone[b]	Infiltration[c]	0.1 – 5.0
Wall[b]	Infiltration	0.1 – 5.0
Bench and trough	Infiltration	0.1 – 5.0

[a] Includes pickhole vent hole chips that are subject to ponding.
[b] Priority brick, block, precast, cast-in-place pipe.
[c] Rainfall induced.

Chemical Grouting. Chemical grouting has been successful in reducing I/I in manhole structures. The process can be a cost-effective option, though it is important to note that grouts do not add to the structural integrity of the manhole. The success of grouts in reducing manhole I/I largely depends on soil and groundwater conditions, injection patterns, gel time and grout mixture, containment of excessive grout migration, and selection of the proper type of grout.

There is a wide range of grouts for pressure application, including acrylamide, acrylate, urethane foam, and urethane gel. The common applications for pressure grouts are for brick manholes with somewhat tight joints, active I/I, structurally sound manholes, cohesive soils with optimal moisture content, and improving and filling voids to stabilize the surrounding soil.

Careful inspection of the grouting work during grouting is essential to ensure adequate coating of the structure exterior. A test program should be performed after grouting to verify proper manhole sealing. Some considerations for high-pressure grouting applications are as follows:

- Differing gel types should be considered for different areas and depths of the manhole structure. This approach has been applied successfully in some projects, although other projects have been successful using only one type of gel for all depths. For projects with varying gel types, urethane and foams have been used in the upper 1.5 m (5 ft) of the manhole, with urethane or acrylamide gels in the lower sections.
- Urethane or acrylamide gels for the lower section.
- Ambient air temperature above 4.5°C (40°F).
- Chemically stable and resistant to acids, alkalis, and organics.
- Controllable reaction times.
- Shrinkage control of 15%.
- Viscosity of approximately 2×10^{-3} Pa (2 cP).
- Constant viscosity during injection period.

Coating Systems. Coating systems have been used to restore manhole structures for several years. In each application, a cementitious material containing Portland cement, finely graded mineral fillers, and chemical additives is applied in one or more layers to the interior of a manhole that has been adequately cleaned and prepared. Most coating systems provide for both mechanical and chemical bonding. The system can be used to coat the entire manhole, including reconstruction of the bench and invert. Coatings are ideally suited for brick structures that show little or no evidence of movement or subsidence, as the coatings have little intrinsic structural qualities in shear and tension.

The most common and successful applications are under the following conditions: brick structure, observed I/I, missing or deteriorated mortar joints, and site conditions that prevent excavation or reconstruction.

Coating systems can be machine or hand applied. Surfaces should be prepared by high-pressure water blasting, using a minimum pressure of 34 500 kPa (5 000 psi) to etch brick and remove defective mortar. All active I/I should be eliminated before coating the manhole. Where necessary, voids should be hand packed with an appropriate patching compound. If the potential for hydrogen sulfide generation exists, the finish coat should be protected with an inorganic liquid polymer product to impregnate and protect the final surface. As an alternative, corrosion-resistant additives can be incorporated to the mix design. The entire process (cleaning, preparation, coating, and cleanup) should be monitored carefully if an independent contractor is performing the work. Postrehabilitation dye-water testing should be performed on a random sample of completed manholes.

There are several coating systems available under various trade names. In general, they are variations of the same basic components and mix designs. The basic coating system and characteristics of each component of the system that produce an effective application for manhole renewal are as follows:

- Patching compounds. Rapid set (20 minutes), self-bonding, high-strength concrete patching mortar, they must be resistant to freeze and thaw environments, with a minimum compressive strength of 27 600 kPa (4 000 psi) over 30 days, a maximum volume change of 0.02%, and contain no calcium chloride, gypsum, lime or high alumina cements.
- Plugging compounds. Used for stopping locations of active infiltration, they must have a set time of 30 to 60 seconds, a mix design conforming to ASTM C150 and ASTM C144, and must exhibit mechanical and active chemical bonding to saturated surfaces.
- Coating compounds. Mix design includes a cementitious pozzolanic mix of Portland cement, chemically active aggregates, and proprietary additives to enhance system performance. They must have a minimum 30-day compressive strength of 20 700 kPa (3 000 psi) and a tensile strength of no less than 10% of compressive strength, with chemical bonds that meet or exceed 1 030 kPa (150 psi) and a coating system that meets or exceeds specifications based on ASTM C495, C496, C293, C596, C666 (Method A), C267, and C321.

If properly applied and under low corrosion-potential conditions, the mechanical and chemical bonding will meet or exceed the 10-year longevity criterion established for most coating systems. A specification should require the contractor to add sulfide-resistant additives to the mix design, and in aggressive corrosive atmospheres, plastic, polymer, or epoxy coatings should be used. The remaining factor that can significantly affect the longevity of the coated structure is creep or long-term differential movement of the manhole structure. Because most coating systems have little shear or tensile strength, cracks will develop.

Structural Linings. Structural rehabilitation of a manhole is any method that totally restores the structural integrity of a manhole through in-place, nondestructive methods. *In situ* rehabilitation methods, such as cast-in-place concrete, have been used in a variety of applications. Reconstruction methods have been limited to manholes with the following conditions: standard manhole dimensions of 1 200- to 1 800-mm (48- to 72-in.) diameter walls, substantial structural degradation, accessible location, substantial project size, and life-cycle cost justification.

Most manhole structures do not warrant structural reconstruction on the sole basis of reducing and controlling I/I, as reconstruction is not cost competitive with coating and pressure grouting based on initial construction cost. Consideration should be given to a life-cycle cost analysis of reconstruction. Structural reconstruction methods include cast-in-place concrete, prefabricated RPM, prefabricated FRP, spiral-wound liner, and CIPP. A typical section through a rehabilitated manhole is shown in Figure 7.17.

General recommendations for structural lining methods are as follows:

- Minimum finished inside wall diameter of 900 mm (36 in.);
- Six-bag (Type I or II) Portland cement mix design;
- Minimum 30-day compressive strength of 24 000 kPa (3 500 psi);
- Minimum wall thickness of 80 mm (3 in.) for manhole depths up to 3 m (10 ft), special analysis beyond 3 m (10 ft); and
- Special cement mixes for corrosion resistance.

Any linings should be designed to withstand external groundwater pressures. Vertical traffic or ground loadings will be carried by the existing manhole structure because it would be difficult to stand the lining on suitable foundations or provide sufficient thickness to carry the loads without restricting the access size. If vertical deterioration is the problem, total replacement may be the most economical solution.

The economics of such rehabilitation depend on such factors as severity of chemical attack or corrosion, location, depth of manhole and water table, number of manholes requiring rehabilitation or replacement, and wastewater flow control measures needed. With precast manholes, it usually is possible to achieve desired results at costs lower than those required for excavation and replacement.

Structural rehabilitation sometimes is not practical, and replacement is necessary. Replacement often is preferable where permafrost or freeze–thaw cycles create special problems.

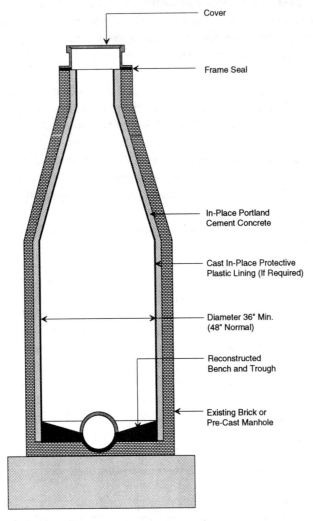

Figure 7.17 Typical section through rehabilitated manhole (in. × 25.40 = mm).

Rehabilitation also should include measures to ensure manhole safety and efficient channel hydraulics. Access ladder rungs and step irons are important safety features and often are suspect. Many communities and wastewater agencies do not install steps in new manholes. Weak rungs should either be replaced with new corrosion-resistant rungs or removed completely.

The efficiency of the present channel also should be evaluated. If the flow is restricted or disturbances are causing extraordinary head losses, repair work should improve hydraulic characteristics. The existing base may have to be partly removed and reconstructed to provide better geometry or surface finish. Flows must be plugged temporarily and quick-setting products used or flows temporarily rerouted while the structure is being repaired. Flexible sleeves can be used to contain flows during repair.

Also important are the entry requirements of maintenance equipment. Cleaning tools, television cameras, and in-line rehabilitation tools, such as grouting packers, all require approximately 600 mm (24 in.) of straight pipe access. The channel should be built accordingly and self-cleaning benching provided.

Corrosion Protection. Manholes subjected to corrosive atmospheres must be protected with a non-cement-type coating. The marketplace offers a variety of plastic, polymer, and epoxy coatings effective in protecting manhole walls from the corrosion of sulfuric acid. Bituminous coatings have not proven to be effective in corrosion control of manholes. The effectiveness of corrosion protection depends on adequate preparation and cleaning of the substrate wall of the manhole and proper application of the coating.

Frame, Cover, and Chimney Rehabilitation. Leakage problems common with manhole frames and covers include surface water entering through the holes in the cover, through the space between the cover and the frame, and subsurface water entering under the manhole frame. Tests consistently have shown that these sources account for a significant portion of manhole leakage.

Ground movement, thermal expansion and contraction of the surrounding pavement, frost heave, and traffic loadings cause the seal between the frame and cover to break and deteriorate, allowing subsurface water to enter the manhole. Entering after running along pavement subgrades, this water washes subgrade material in with it, causing pavement settling around the manhole.

Manhole covers can be sealed by replacing them with new watertight covers, by sealing existing covers through the use of rubber cover gaskets and rubber vent and pick hole plugs, or by installing watertight inserts under the existing manhole covers (Figure 7.18). Selection of watertight inserts (plastic, fiber glass, stainless steel) must consider installation condition's (nontraffic, heavy traffic) effect on the required strength and durability of the insert.

The manhole frame-chimney joint area can be sealed internally without excavation when frame alignment and chimney condition permit, or internally or externally when realignment or replacement of the frame or reconstruction of the chimney or cone requires excavation. This sealing can be achieved by installing a flexible manufactured seal or by applying a flexible material to either the surface of the chimney and frame or between the adjusting rings and under the frame.

The method used must both be watertight and have the flexibility to allow for the repeated movement of the frame throughout the manhole's design life.

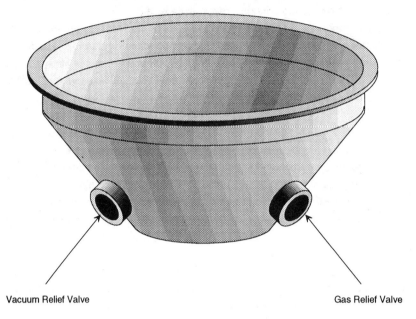

Vacuum Relief Valve   Gas Relief Valve

Figure 7.18   Polyethylene manhole insert.

SERVICE CONNECTION REHABILITATION

Service connections are pipelines that branch off the sewer main and connect building sewers to the public sewer main. They may be as small as 100 mm (4 in.) in diameter, but normally range from 4.5 to 30 m (15 to 100 ft) or more in length.

Service connections are built with any one of several products, and usually are laid at a minimum self-cleansing grade from the building to the immediate vicinity of the main sewer. At this point the grade may change abruptly for the line to descend to the main sewer. Service connections normally enter sewer mains at angles ranging from 30 to 90 deg from the axial flow direction and at vertical angles ranging from 0 to 90 deg. In some developments, the same trench is used to route potable water service connections and the sewer service connection line. Consequently, any leaks in the potable water line can enter the sewer service connection line if it is not watertight.

Construction and maintenance of service connections is complicated because separate government agencies usually have jurisdiction over different portions of sewers. The connection between a building's plumbing and drain system and the property line often is considered an extension of the instructure facilities, thus ordinarily is installed under plumbing or building codes and tested and approved by plumbing officials or building inspectors. The sewer section between the property line and the street sewer, including the sewer main connection, usually is installed under sewer use rules. Public works or sewer officials are responsible for inspection and approval.

Industrial waste connections may be exceptions. The construction of many entire industrial service connections is supervised by collection system officials because of possible effects of such wastes on sewer structures and treatment facilities.

For many years, the effect of leaking service connections on the collection system and treatment facilities was considered insignificant. It was assumed that most service connections were above the water table and subject to leakage only during periods of excessive rainfall or exceedingly high groundwater levels.

The need to repair service connections now is widely recognized, and I/I studies have demonstrated the impact of neglected building sewers on collection systems and treatment facilities. Research studies sponsored by U.S. EPA (Grace and Burkhard, 1991, and Reed, 1987 and 1988) indicate that a significant percentage of I/I often is caused by defects in service connections, including cracked, broken, or open-jointed pipes, which can allow storm-induced infiltration. Service connections may also transport water from inflow sources such as roof drains, cellar and foundation drains, basement or subcellar sump pumps, and clean water from commercial and industrial effluent lines. Service connections also can be infiltrated by water migrating after sewer mains have been repaired.

The potential for infiltration from service connections depends on the number of connections and total length of the connection lines. Collection sewers in urban areas can contain multiple service connections. Furthermore, the total length of service connections often is equal to or greater than the sewer main length. For example, lots with a 15-m (50-ft) street frontage usually have four service connections per side of the street, or eight per 60 m (200 ft) of block length. If the average service connection is 8 m (25 ft) to the street line, the total length of these lines will be equal to the length of the sewer main.

Where high-density housing construction is common, 75% of the infiltration can result from service connections.

REHABILITATION METHODS. Some rehabilitation methods used for larger main sewers also apply for house connections and laterals. These include chemical grouting, CIPP lining, deformed pipe lining, and pipe bursting. The reduction in capacity imposed by sliplining methods makes such rehabilitation an unlikely option in smaller pipelines.

Access to service connections often can be limited. This can make rehabilitation difficult, along with the additional problems of tree roots and landscaping over the sewers in suburban areas. These problems have a greater impact on conventional replacement systems.

Chemical Grouting. Three chemical grouting methods are available for sealing building sewers: the pump full method; the sewer sausage method (patented process); and the camera-packer method (patented process).

The pump full method involves injecting a chemical grout through a conventional sealing packer from the sewer main up the service connection to an installed plug. As the grout is pumped under pressure, it is forced through the pipe faults into the surrounding soil, where a seal is formed after the gel has set. After sealing, excess grout is augered from the building sewer and the sewer is returned to service.

The sewer sausage method is similar to the pump full method in that it requires access to the building sewer, the use of a camera-packer unit in the sewer main, and the injection of grout from the sewer main up the service connection to seal the pipeline. The primary difference is that a tube is inserted to the service connection before sealing to reduce the quantity of grout used and to reduce the amount of cleaning required after sealing. The grout is pumped under pressure around the tube, up the service connection, and through any pipe faults into the surrounding soil, where the seals are formed after the gel sets.

Unlike the other methods, the camera-packer method does not require placing equipment in the sewer main. It also differs in concept because only faults seen through a television camera are repaired. A miniature television camera and a specialized sealing packer are inserted to the service connection. Using a tow line previously floated from the service connection access to the downstream manhole of the sewer main, the camera-packer unit is pulled into the service connection and then slowly pulled back out, repairing faults that are seen through the television camera.

The deepest leaking joints are sealed first. Joints and cracks are sealed in a manner similar to the conventional methods used for sealing joints in sewer mains. When repairs have been completed, the equipment is removed and the service connection returned to service. The costs for this method vary, depending on the difficulties encountered during repairs. When estimating costs, allowances should be made for such things as difficult site access and excavation dewatering.

Cured-in-Place Pipe Lining. The technology for sealing lines as small as 100 mm (4 in.) by CIPP lining is in common use. As with sewer mains and laterals, CIPP lining should reduce infiltration and improve the structural integrity of the existing pipeline.

The steps for lining a service connection using the CIPP process are similar to those for lining a sewer main, though an access point requiring excavation usually is needed on the upstream side of the service connection line. Another variation from sewer main installations is the use of a special pressure chamber to provide pressure to invert the fabric material through the service pipeline. The fabric is terminated at the entrance of the sewer main, instead of at a downstream manhole.

After curing, the downstream end of the liner is opened by excavation or a remotely controlled cutting device placed in the sewer main. The upstream end is trimmed and the newly lined pipe connected to the rest of the existing service connection line, restoring sewer service. The excavated soil is then replaced and all equipment removed.

OTHER MEASURES. The methods discussed to rehabilitate service connection lines should reduce infiltration resulting from high groundwater levels and rainstorms. Efforts also should be made to remove any sources of inflow related to service connections. Inflow sources connected to service connection lines usually are located on private property. Public awareness or public relations programs often are needed. Such programs are intended to persuade property owners (without threat of legal consequences) to make the needed repairs to help correct a community problem. Table 7.10 lists typical types of inflow sources found on private property and possible rehabilitation measures.

Table 7.10   Miscellaneous private property rehabilitation measures.

Inflow source	Possible rehabilitation measure
Connected downspout	Plug service connection opening and redirect downspout
Connected storm sump	Repipe to grade
Connected storm sump with diverter valve	Remove diverter valve
Defective or broken cleanout	Repair or replace as necessary
Connected area drain	Disconnect drain and install new sump pump
Connected crawlspace drain	Seal drain and install new sump pump
Connected foundation drain	Disconnect drain and install new sump pump

REFERENCES

Briassoulis, D., *et al.* (1989) Static and Dynamic Effects of the Pipe Insertion Machine (PIM) Technique. U.S. Army, Corps Eng., Constr. Eng. Res. Lab. Tech. Manuscript N-89/07.

Decker, C., and Larson, J. (1988) California Sewer Replacement Experience Using Pipe Bursting (PIM) Technology. No-Dig '88, Washington, D.C.

Grace, J.C., and Burkhard, M. (1991) Structural Spray-On Linings: A Radical New Approach. North America No-Dig '91, Kansas City, Mo.

Horne, J.B., et al. (1987) Rolldown. No-Dig '87, London, Eng.

Information and Guidance Notes 4-12-04, 4-32-05, 4-34-02, 4-34-04, and 4-24-05 (1986). Water Authorities Assoc., Sewers and Water Mains Comm. Mater. and Stand., Eng.

Kirby, M.J. (1991) Extending the Capabilities of Fluid Jet Directional Drilling. North America No-Dig '91, Kansas City, Mo.

Knott, G.E. (1989) A Review of Non-Man Entry Sewer Cleaning Practices. Foundation for Water Res., Rep. FR0030, Eng.

Knott, G.E. (1990) Non-Man Entry Sewer Cleaning: Report on Equipment Performance. Water Res. Cent., Report UM 1066, Eng.

Ledoux, P., and Catha, S.C. (1988) Advantages and Applications of the U-Liner System. No-Dig '88, Washington, D.C.

Lee, L. (1991) Design and Specification of Slipliner and Tunnel Liner Grouting Systems. North America No-Dig '91, Kansas City, Mo.

Leech, G., and Reed, K. (1989) Observation and Assessment of the Disturbance Caused by Displacement Methods of Trenchless Construction. No-Dig '89, London, Eng.

Manual on Buried PVC Pipe Performance (1982). Wavin R & D, Neth.

Menzel, S.W.O. (1988) Development and Case Histories of Spirally Wound Liners. No-Dig '88, Washington, D.C.

Nelson, R.E. (1991) Manhole Evaluation and Rehabilitation. Rep. Am. Soc. Chem. Eng. Pipeline Infrastructure Committee, Manhole Task Committee, New York, N.Y.

Reed, K. (1987) The Application of Moling Techniques in the Water Industry. No-Dig '87, London, Eng.

Reed, K. (1988). An Overview of Pipe Replacement Using Moling Techniques in the United Kingdom. No-Dig '88, Washington, D.C.

Reynolds, J.M., and Szczupak, J.R. (1987) Directional Drilling Experiences. No-Dig '87, London, Eng.

Sewerage Rehabilitation Manual (1986) (Addendum in 1990). 2nd Ed., Water Res. Cent., Eng., U.K.

Sewer Segment Flow Monitoring Study to Establish In-Service Manning Roughness Coefficients (1990). Sverdrup Corporation and Southeast Environ. Serv. Insituform of North America, Inc.

Steketee, C.H. (1988) Nu-Pipe, A New Method of Rebuilding Underground Pipes. No-Dig '88, Washington, D.C.

Thomson, J. (1991) The Potential for Microtunnelling in North America. North America No-Dig '91, Kansas City, Mo.

Wade, M.G. (1990) Assessment, Evaluation, and Rehabilitation of Manhole Structures: Innovative Methods. Paper presented at 63rd Annu. Conf. Water Pollut. Control Fed., Washington, D.C.

Water Research Centre (1988) A Preliminary Assessment of Microtunnelling and Other Trenchless Systems for the U.K. Market. Eng. Rep. ER318E, Eng.

Chapter 8
Pipeline Rehabilitation Materials

- 200 Sliplining
- 200   Continuous Pipe
- 200   High-Density Polyethylene Pipe
- 200   Polybutylene Pipe
- 201   Short Pipe
- 201   High-Density Polyethylene Pipe
- 201   Polyvinyl Chloride Pipe
- 202   Fiber Glass Pipe
- 203   Ductile Iron Pipe
- 203   Steel Pipe
- 203 Cured-in-Place Pipe
- 204 Deformed Pipe
- 204   U-Liner Pipe
- 204   Nu-Pipe
- 204   Rolldown
- 205 Spiral-Wound Pipe
- 206 Segmental Linings
- 206   Fiber-Glass-Reinforced Cement
- 206   Fiber-Glass-Reinforced Plastic
- 206   Reinforced Plastic Mortar
- 206   Polyethylene
- 206   Polyvinyl Chloride
- 207   Welded Steel
- 207 Coatings
- 207 Replacement Materials
- 207 Maintenance and Repair Materials
- 207   Root Control
- 207   Chemical Grouting
- 208   Acrylamide Grout
- 209   Acrylic Grout
- 209   Acrylate Grout
- 209   Urethane Grout
- 210   Urethane Foam
- 210 Reference

New materials for sewer pipeline rehabilitation continually are being developed to rehabilitate sewer systems and appurtenances. Sewer system rehabilitation may require different materials than those used for new sewer construction.

The growth of pipeline rehabilitation materials has been phenomenal, and the user must properly review and evaluate them. In anticipation of field and installation uncertainties, the long-term performance of rehabilitation materials against anticipated, installed, or *in situ* environmental conditions should be evaluated. Material parameters, such as installation stresses, long-term hydrostatic buckling pressures, and other *in situ* structural-resistant requirements, must be substantiated by the supplier through long-term testing programs and provided to the user. The materials should be certified as meeting national, manufacturing, and project standards.

Factors to be considered in selecting sewer rehabilitation materials include long-term flow characteristics, durability, installation procedures, availability, long-term modulus value (50 years), and cost. Material selection should be based on specific requirements.

When an existing sewer is to be lined, available materials include sliplining, cured-in-place pipe, deformed pipe, spiral-wound pipe, segmental linings, coatings, replacement materials, and maintenance and repair materials.

SLIPLINING

CONTINUOUS PIPE. High-Density Polyethylene Pipe. High-density polyethylene (HDPE) pipe materials should conform to the following standards:

- Polyethylene (PE) Plastic Pipe Based on Outside Diameter, ASTM F-714;
- Polyethylene Plastics Pipe and Fitting Materials, ASTM D-3350; and
- Polyethylene Plastics Molding and Extrusion Materials, ASTM D-1248.

The polyethylene (PE) cell classification is 345434C. Additives and fillers, including stabilizers, antioxidants, and lubricants, may not exceed 5 parts per 100 parts by weight of PE resin in the compound.

The PE pipe is manufactured as low-, medium-, and high-density resins. High-density is used for sewer sliplining. Structural characteristics are a function of wall thickness, and high-density compounds are rigid, hard, strong, tough, and corrosion resistant.

Because of the low modulus of elasticity, careful design considerations must be given to buckling. The PE pipe is characterized by the standard dimensional ratio, and pipe grades are classified by environmental stress-cracking resistance. The ASTM D-3350 standard is normally referred to for pipe cell classification for material properties. The cell classification for sliplining pipe is 345434C. The primary properties of concern for PE sliplining pipe are listed in Table 8.1.

Polybutylene Pipe. Polybutylene (PB) pipe materials should conform to the following standards:

Table 8.1 Primary properties of concern for polyethylene sliplining pipe.

Property	ASTM test method	Value
Density, g/cm^3	D-1505	0.941–0.955
Melt index-condition E	D-1238	< 0.15
Flexural modulus, psi[a]	D-790	110 000 – 160 000
Tensile strength at yield, psi	D-1693	3 000 – < 3 500
Environmental stress cracking resistance	D-1693	
Test condition		A
Test duration, hours		192
Failure, maximum %		0
Hydrostatic design basis, psi	D-2837	1 600
Color and ultraviolet stabilizer		C
		2% carbon black, min.

[a] psi × 6.895 = kPa.

- Large Diameter Polybutylene Plastic Pipe, ASTM F-809; and
- Polybutylene (PB) Plastics Molding and Extrusion Materials, ASTM D-2581.

Polybutylene is a material with a stiffness resembling that of low-density polyethylene and long-term strength greater than that of HDPE. It also is less affected by increasing temperature than PE. While most PEs have an upper temperature limit of approximately 60°C (140°F), the limit for PB is nearly 93°C (200°F).

SHORT PIPE. High-Density Polyethylene Pipe. High-density polyethylene profile wall pipe materials should conform to the following standard:

- Polyethylene Large Diameter Profile Wall Sewer and Drain Pipe, ASTM F-894.

The primary properties of concern and the material standards that should be met are the same as those for solid wall HDPE pipe. The pipe should be tested for pipe stiffness in accordance with ASTM D-2412.

Polyvinyl Chloride Pipe. Polyvinyl chloride (PVC) pipe materials should conform to the following standards:

- Type PSM Poly (Vinyl Chloride) (PVC) Sewer Pipe and Fittings, ASTM D-3034;
- Poly (Vinyl Chloride) (PVC) Large-Diameter Plastic Gravity Sewer Pipe and Fittings, ASTM F-679; and

- Rigid Poly (Vinyl Chloride) (PVC) Compounds and Chlorinated Poly (Vinyl Chloride) Compounds, ASTM D-1784.

Polyvinyl chloride pipe should be extruded of plastic materials having a cell classification of 12454-B, 13364-A, or 13364-B, as defined in ASTM D-1784. Additives and fillers, including stabilizers, antioxidants, lubricants, and colorants, are to be 10 parts or less per 100 parts by weight of PVC resin in the compound. The primary properties of concern for PVC sliplining pipe are listed in Table 8.2.

Fiber Glass Pipe. Reinforced plastic mortar and fiber-glass-reinforced plastic pipe materials should conform to the following standards:

- Fiber Glass (Glass-Fiber-Reinforced Thermosetting Resin) Sewer Pipe, ASTM D-3262; and
- Chemical Resistance of Fiber Glass (Glass-Fiber-Reinforced Thermosetting-Resin) Pipe in a Deflected Condition, ASTM D-3681.

Fiber glass pipe with or without sand mortar can be manufactured with polyester, vinylester, or epoxy resin. The pipe must conform to the strain-corrosion testing requirements provided in ASTM D-3681. The pipes can be manufactured with pipe stiffness 125, 250 or 500 kPa (18, 36, or 72 psi) and should be tested in accordance with ASTM D-2412.

Table 8.2 Primary properties of concern for polyvinyl chloride sliplining pipe.

Property	ASTM test method	Value[a]
Base resin		Polyvinyl chloride homo-polymer (1) (2) (3)
Impact strength, ft–lb/in. of notch[b]	D-256 Method A	0.65 (1) (2) 1.5 (3)
Tensile strength, minimum, psi[c]	D-638	6 000 (2) (3) 7 000 (1)
Modulus of elasticity in tension, minimum, psi	D-638	400 000 (1) 440 000 (2) (3)
Deflection temperature under load, minimum, °F[d]	D-648	158 (1) (2) (3)
Chemical resistance	D-543	(1) (3) (2)

[a] Cell classification: (1) 12454—B; (2) 13364—A; and (3) 13364—B.
[b] ft–lb/in. × 1.659 = J/m.
[c] psi × 6.895 = kPa.
[d] 0.555 6 (°F − 32) = °C.

Ductile Iron Pipe. Ductile iron pipe should conform to the following standards:

- Thickness Design of Ductile Iron Pipe, AWWA C-150;
- Ductile Iron Pipe, Centrifugally Cast in Metal Molds or Sand-Lined Molds, for Water or Other Liquids, AWWA C-151; and
- Cement-Mortar Lining for Ductile-Iron and Grey-Iron Pipe and Fittings for Water, AWWA C-104.

Where specific conditions make pipe protection necessary, such as high amounts of hydrogen sulfide or sulfuric acid, a polyethylene or other suitable liner can be provided. A virgin PE liner conforming to ASTM C-1248 compounded with filler materials, including pigment, should be used. The PE liner material should be 40 mils (1 mm) in thickness and heat bonded to the inside of the pipe over the full length of corrosive exposure.

Steel Pipe. Steel pipe should conform to the following standards:

- Steel Water Pipe 150 mm (6 in.) and Larger, AWWA C-200;
- Coal-Tar Protective Coatings and Linings for Steel Water Pipelines—Enamel and Tape—Hot Applied, AWWA C-203; and
- Fusion-Bonded Epoxy Coating for the Interior and Exterior of Steel Water Pipelines, AWWA C-213.

Welded steel pipe is made by butt or offset butt electrically welded straight or spiral seam cylinders, fabricated from plates or sheets. The plates or sheets comply with the physical and chemical requirements of ASTM A-570 or ASTM A-283. Though determined by the design, wall thickness usually will be 10 to 15 mm (0.375 to 0.5 in.), depending on the corrosion potential. The finished in-place liners will be pressure grouted and lined with 15- to 20-mm (0.5- to 0.75-in.) cement mortar having corrosion-resistant additives. Where specific conditions make conventional linings marginal, such as large amounts of hydrogen sulfide or sulfuric acid, it may be necessary to use stainless steel series 300 material.

CURED-IN-PLACE PIPE

Cured-in-place pipe (CIPP) should conform to the following standard:

- Rehabilitation of Existing Pipelines and Conduits by the Inversion and Curing of a Resin Impregnated Tube, ASTM F-1216.

The lining consists of a tube of polyester material saturated with a thermosetting polyester, epoxy, or vinylester resin. Thickness is increased by adding additional layers of the polyester material, which can be felt or a nonwoven or woven fabric. The woven fabric increases the tensile and flexural properties of the lining, and the epoxy resins increase the modulus of elasticity of the cured pipe. A final layer of polyurethane is bonded to the fabric to prevent washout of the resin and contains the water or steam used for inversion or inflation and curing. The liner tube can be impregnated with resin at the factory and immediately transported to the site in refrigerated containment or can be impregnated on site. The primary properties of concern for CIPP are listed in Table 8.3.

Table 8.3 Primary properties of concern for cured-in-place pipe.

Property	ASTM test method	Value[a]	Polyester[b]
Tensile strength, psi[c]	D-638	4 000	3 000
Flexural strength, psi	D-790	5 000	4 500
Tensile modulus, psi	D-638	250 000	200 000
Flexural modulus, psi	D-790	300 000	250 000

[a] Values listed are considered initial minimums as defined in ASTM D-638 and D-790. Long-term values are defined as 50 year and are determined by ASTM D-2290 test method.
[b] Enhanced polyester resins are available providing higher physical properties.
[c] psi × 6.895 = kPa.

DEFORMED PIPE

U-LINER PIPE. U-liner pipe, deformed into a U-shape, is manufactured HDPE. This deformed HDPE pipe references ASTM D-3350 for pipe cell classification for material properties. The cell classification for PE deformed pipe is 345434. These properties are identical to PE sliplining pipe previously provided in this chapter, except the color and ultraviolet stabilizer is normally white (titanium dioxide). This feature permits the locating of service laterals with closed-circuit television (CCTV) for *in situ* connection.

NU-PIPE. Nu-pipe folded into a U-shape is manufactured PVC pipe. This folded PVC pipe references ASTM D-1784 for pipe cell classification for material properties. The cell classification for PVC folded pipe is 12334-B. The primary properties of concern for PVC deformed pipe are as in Table 8.4.

ROLLDOWN. Rolldown systems are deformed during and reformed after installation. They are manufactured HDPE with pipes that reference ASTM D-3350 for pipe cell classification for material properties. Their cell classifi-

Table 8.4 Primary properties of concern for polyvinyl chloride deformed pipe.

Property	ASTM test method	Value
Base resin		Polyvinyl chloride homo-polymer
Impact strength, ft–lb/in. of notch[a]	ASTM D-256 Method A	0.65
Tensile strength, minimum, psi[b]	ASTM D-638	6 000
Modulus of elasticity in tension, minimum, psi	ASTM D-638	320 000
Deflection temperature under load, minimum, °F [c]	ASTM D-648	158
Chemical resistance	ASTM D-543	B

[a] ft–lb/in. × 1.659 = J/m.
[b] psi × 6.895 = kPa.
[c] 0.555 6 (°F–32) = °C.

cation is 345434. These properties are identical to PE sliplining pipe, except the color and ultraviolet stabilizer are normally white (titanium dioxide). This feature permits locating service laterals with CCTV for *in situ* connection.

SPIRAL-WOUND PIPE

Spiral-wound pipe is PVC profile former strips and joiner strips manufactured by extrusion. The PVC material properties are identical to the PVC sliplining pipe previously provided in this chapter. The minimum thicknesses of the plastic strips and panels are listed in Table 8.5.

Table 8.5 Dimensions of polyvinyl chloride strips and panels.

Nominal diameter, in.[a]	Minimum thickness		Minimum profile height, in.
	Former strip, in.	Joiner strip, in.	
8–12	0.025	0.025	0.192
15–18	0.030	0.031	0.242
24–36	0.045	0.058	0.480
30–72[a]	0.060	—	0.488

[a] in. × 25.4 = mm.

SEGMENTAL LININGS

FIBER-GLASS-REINFORCED CEMENT. Fiber-glass-reinforced cement linings are prefabricated panels normally 10 mm (0.375 in.) thick. These linings are composed of Portland cement, fine sand, and chopped fiber glass rovings. The fiber glass must have a surface finish compatible with the high alkaline cement environment. These linings have high mechanical and impact strengths, with negligible absorption and permeability features.

FIBER-GLASS-REINFORCED PLASTIC. Fiber-glass-reinforced plastic (FRP) linings are approximately 15 mm (0.5 in.) thick but can vary, depending on the material properties desired. They are composed of thermosetting plastic resin and chopped fiber glass rovings. The fiber glass reinforcement is of the acid-corrosion-resistant variety. The resins employed normally are bisphenol A, vinylester, or isophthalic-acid-resistant types. These liners are highly resistant to abrasion, with negligible absorption and permeability features.

REINFORCED PLASTIC MORTAR. Though, similar to FRP linings, reinforced plastic mortar (RPM) liners have an added fine sand and resin mortar in the matrix. Also, the RPM design may use continuous rovings.

POLYETHYLENE. Polyethylene sheet liners can be manufactured in thicknesses of 40 to 240 mils (1 to 6 mm). The sheet thickness should be selected for the anticipated temperature range, diameter of the pipe, and degree of circumference to be covered. The HDPE sheets must be free of extractable plasticizers or copolymers, maintaining physical properties with low-level shrinkage or swelling from normal wastewater conditions.

POLYVINYL CHLORIDE. Polyvinyl chloride linings are composed of high-molecular-weight PVC resin combined with chemical-resistant pigments and plasticizers. The PVC liner is extruded of plastic materials having cell classifications identical to the PVC sliplining pipe previously provided in this chapter.

The linings normally are 65 mils (1.7 mm) thick, with a minimum elongation of approximately 200% (ASTM D-412) and a tensile strength of 15 200 kPa (2 200 psi [ASTM D-638]). The Shore Durometer hardness is approximately 50 at ambient temperature (ASTM D-2240). The standard sheets are 1.2 by 2.4 m (4 by 8 ft), having a 10-mm (0.375-in.) high tee profile section, spaced 65 mm (2.5 in.) apart circumferentially and running longitudinally. Water path sections are provided longitudinally to permit groundwater passage.

WELDED STEEL. Welded steel lining placed in the pipeline in segments is identical to the welded steel pipe sliplining previously discussed in this chapter.

Coatings

Reinforced shotcrete (gunite) and cast-in-place concrete are discussed in Chapter 7.

Replacement Materials

The materials used in trenchless technology include HDPE, PVC, RPM, and FRP. Materials used in conventional replacement are provided in *Gravity Sanitary Sewer Design and Construction* (WPCF, 1982).

Maintenance and Repair Materials

ROOT CONTROL. Chemical root treatment (foaming method) is intended to kill roots and inhibit regrowth without damaging trees or plants, the environment, or the wastewater treatment plant process. The chemical root treatment must be registered with the U.S. Environmental Protection Agency (U.S. EPA) and labeled for use in sewer pipelines.

The active ingredient for killing roots is usually a nonsystemic herbicide that kills roots at low concentrations. The active ingredient must be detoxified by natural processes following its use. The common composition of such a chemical is 24.25% sodium methyldithiocarbonate (anhydrous), 1.77% dichlobenil (2,6-dichlorobenzonitrile), and 73.98% inert ingredients.

The sewer pipe capacity is 0.1 L of foam per linear mm (0.7 gal per linear in.) for a 100-mm (4-in.)-diam pipe, 5.7 L for a 150-mm (1.5 gal for a 6-in.) pipe, 9.5 L for a 200-mm (2.5 gal for an 8-in.) pipe, 15 L for a 250-mm (4.0 gal for a 10-in.) pipe.

CHEMICAL GROUTING. There are several types of chemical grouts, categorized as either gel or foam. With each of the grouts there are many types of additives, such as initiators, activators, inhibitors, and various fillers.

The general grout formulations are chemicals and water. When there is groundwater present, normal practice dictates that a higher concentration of chemicals be used because of the dilution potential. Because of soil and moisture variability, formulating the correct mixture depends on trial and error rather than scientific principles.

The various parameters that affect performance are viscosity control, gel time variables, temperature, pH, entrained oxygen in the solution, contact with particular metals, ultraviolet rays, mineral salts, the velocity of groundwater flows, capabilities of placement equipment, and other soil and water conditions. The properties of the grout also vary in appearance, solubility, swelling and shrinkage, corrosiveness, stability, and strength. Because various grout additives affect viscosity, density, color, strength, and shrinkage, for proper formulation environmental conditions must be considered on a case-by-case basis.

Another aspect of effective grouting is the proper use of the equipment and process, that is, the grout packer, pumps, tanks, formulation, mixing, and application. Premixing of the final grout mixture is conducted in two separate tanks—the grout tank and the catalyst tank. The grout tank contains water, grout, and buffer, and the catalyst tank contains water, oxidizer catalyst, and fillers.

The most commonly used gel grouts have acrylamide, acrylic, acrylate, or urethane bases. All the gels are resistant to most chemicals found in sewer pipelines and produce a gel–soil mixture susceptible to shrinkage cracking. All except the urethane-base gels are susceptible to dehydration. These deficiencies can be reduced by using chemical additives.

In addition to chemical differences in composition, there are other important differences. Acrylamide-base gel is significantly more toxic than the others, though grout toxicities are of concern only during handling and placement or installation. Nontoxic, urethane-based gels are U.S. EPA approved for potable water pipelines and use water as the catalyst, unlike the other gels which use other chemicals. Therefore, the urethane gels must be free of additional water contamination for proper curing.

According to U.S. EPA, acrylamide grout contains carcinogens. A new, nonhazardous chemical grout is now being used in some areas in the U.S. and likely will be the choice of many current acrylamide grout users.

Acrylamide Grout. This type of grout is a mixture of three or more water soluble chemicals. The base chemical in the mixture, acrylamide normally is at least 10% of the total mixture weight.

Catalysts, such as triethanolamine (TEA) or T+ and ammonium persulfate (AP), are part of the mixture. The T+ is normally 1%, and the AP 0.5% of the total mixture. The catalyst percentages are increased to shorten gel times. Also, gel times will be slower if the temperature of the grout mixture decreases. Gel time is reduced by approximately half for each 5.6°C (10°F) increase in temperature. The controlled reaction time can be varied from approximately 10 seconds to 1 hour. Where high or flowing groundwater is encountered, the desired reaction gel time normally is set faster to prevent diluting of the grout mix.

The catalyst T+ acts as a buffer, and the AP is a granular, strong, oxidizing initiator chemical. A shrinkage-control agent sometimes is added to the mixture for protection against freezing temperatures or dehydrating conditions required for the gel. One or more catalytic aids are also added to reduce the amounts of other catalysts required to achieve a specific gel time. There are chemicals that enhance compressive and tensile strengths and elongation properties. Potassium ferricyanide behaves as an inhibitor and can be used to extend the gel time. A filler, such as diatomaceous earth, can be added to improve shrinkage characteristics, also shortening gel time.

Additives for external grouting are diatomaceous earth, silica flour, sawdust, or bentonite for reducing shrinkage in large void filling. The use of Portland cement for external grouting reduces the gel time but forms a more rigid mass. The additives increase the viscosity and prevent it from being filtered out of the chemical sealant. Certain chemical additives will maintain a true single-phase chemical grout, whereas particulate additives transform the grout to a two-phase system. Depending on the soil grain sizes and other conditions, the additive is normally 5 to 30% by weight of the mixture. Where fine soils are present, the use of fillers should be less than 5%. The amount of water can vary for the mixture but is normally 75 to 90%. Without fillers, the grout mixture's viscosity is similar to that of water.

Acrylic Grout. These types of grout are water solutions of several types of acrylic resins for different applications. After being mixed with catalysts, a cohesive gel is formed. The set time can be closely controlled from a few seconds for flowing water conditions to several hours for normal conditions.

Acrylic grouts are good to use in sewer pipe joints, manholes, and structures. They have a viscosity before gelation similar to water and tend to swell in water, allowing a water-tight seal. The standard formulation for acrylic gel is 167 L (44 gal) water, 53 L (14 gal) acrylic grout, 1.9 L (0.5 gal) or 1% TEA, and 2.3 kg (5 lb) sodium persulfate.

Acrylate Grout. These are similar to acrylic grouts. The standard formulation, by weight, for acrylate gel is 61% water, 35% acrylate grout, 2% TEA, and 2% AP. Where shrinkage control is important, the approximate formulation, by weight, would be 56% water, 35% acrylate grout, 2% TEA, 2% ethylene glycol, and 5% shrinkage inhibitor. Acrylate grout is in a water solution at 40% concentration.

Urethane Grout. This type of grout is a solution of a prepolymer that cures on reaction with water. During the reaction the gel remains hydrophilic, absorbing water and holding it within a cured gel mass. After being cured, the resultant gel is resistant to the passage of water. Because the prepolymer is cured by water, premature contamination by other water must be avoided during application.

The formulation is water sensitive, and various water-to-grout ratios provide various strengths. The ratio needed to provide a strong gel ranges from 5:1 to 15:1, by volume. Ratios less than this will produce a foam reaction, and greater ratios will produce a weak gel.

A sample formulation would involve an 8:1 ratio of compounded material for pipe joint grouting; for example, a water side tank might have 151 L (151 kg) (40 gal [333 lb]) water, 38 L (36 kg) (10 gal [80 lb]) shrink control, and 23 kg (50 lb) diatomaceous earth.

The normal procedure is that the water is added to a 190-L (50-gal) tank, the shrink control is added, and it is thoroughly mixed. A filler may be added, with volumetric adjustments made because of tank capacity. The mixture containing filler is continuously agitated. Under application conditions, 30 to 38 L (8 to 10 gal) of the mixture are combined with each 3.8 L (1 gal) of urethane chemical for grouting.

Urethane Foam. Foam instead of grout is used primarily to stop infiltration to manholes. These leaks occur through cracks in the foundation, base, or wall; joints formed by the base, corbel, or upper frame interfaces; or pipe penetrations to the manhole wall. Grout injections are placed into predrilled holes and the grout is injected under pressure. When cured, it forms a flexible gasket or plug in the leakage path. When mixed with an equal amount of water, the grout expands and quickly cures to a tough, flexible, closed-cell rubber. In some applications, the material is used without premixing with water; however, it will eventually require an equal amount of water for complete curing.

*R*EFERENCE

Water Pollution Control Federation (1982) *Gravity Sanitary Sewer Design and Construction.* Manual of Practice No. FD-5, Washington, D.C.; Manual of Engineering Practice No. 60, Am. Soc. Civ. Eng., New York, N.Y.

Chapter 9
Quality Assurance

213 Setting the Objectives of Sewer Rehabilitation
213 Infiltration and Inflow Control Objectives
213 Selecting the Design Condition
213 Allocating System Components for Quantified Infiltration and Inflow
214 Estimating the Effectiveness of Rehabilitation for Infiltration and Inflow Reduction
215 Structural Rehabilitation Objectives
215 Construction-Phase Quality Assurance
215 Material Quality Control
216 Existing Sewer Conditions
216 Safety
216 Preparation of Sewers
216 Installation
216 Annular Grouting
217 Contractor Payment
217 Construction Inspection
217 Measuring the Effectiveness of Infiltration and Inflow Control
218 Methods for Normalizing Flow Data
218 Infiltration Simulation
219 Inflow Simulation
219 Sanitary Flow Simulation
219 Storm Inflow Comparison
220 Flow-Monitoring Considerations
220 Studies of Rehabilitation Effectiveness for Infiltration and Inflow Control
221 Measuring the Effectiveness of Rehabilitation for Structural Integrity
221 Specific Repair Effectiveness
221 Overall System Integrity
221 Expected Effectiveness of Special Sewer Rehabilitation Methods
221 Sewer Replacement
222 Sewer Relining
222 Lining and Sliplining
223 Inversion Lining
223 Sewer Sealing
224 Service Lateral Rehabilitation
224 Inflow Control
224 Manholes
225 Catch Basins
225 Roof Drains
225 Other
225 Overall Sewer System Effectiveness of Infiltration and Inflow Control
226 Continuing Sewer Maintenance
227 Data Needs for Budgeting Preventive Maintenance
228 Sewer Maintenance Activities
228 Annual Reporting
228 References

The objective of sewer rehabilitation is to maintain the overall viability of a conveyance system. This is done in three ways: (1) by ensuring its structural integrity, (2) limiting the loss of conveyance and wastewater treatment capacity through reducing infiltration and inflow (I/I), and (3) limiting the potential for groundwater contamination through controlling exfiltration from the pipe network.

The rehabilitation method will depend on the rehabilitation objectives. While most structural rehabilitation methods also reduce rates of infiltration, many infiltration control methods have little or no impact on structural integrity.

Over the last two decades, increased emphasis has been placed on sewer rehabilitation in an attempt to reduce I/I identified as economically excessive, thus economical to remove. There is more emphasis on structural rehabilitation efforts that coincides with more recent public awareness related to the implications of deterioration in the nation's core infrastructure.

This discussion of quality assurance includes information on both the ability of rehabilitation methods to reduce I/I and to improve or maintain structural integrity. Information on the effectiveness of a number of inflow control methods for flow reduction also is included.

Quality assurance of rehabilitation efforts should be focused on achieving both the objectives of the specific rehabilitation effort and the overall system objective, which reflects the sum of the individual efforts.

The underachievement of many rehabilitation efforts in reducing I/I is related to the failure to achieve overall system objectives, not necessarily the failure of the individual rehabilitation effort. Thus, sewer rehabilitation investigations are forensic in nature, and it is difficult to achieve a full understanding of the actual sources of I/I and subsequent systemwide effects when leaks in a particular element of the overall system are plugged.

Structural emphasis programs have been more successful. From a community service perspective, the effectiveness of structurally focused rehabilitation should also be measured on a systemwide perspective by eliminating unscheduled structural repairs.

The following sections present data and discussions on (1) quality control procedures used to ensure the effectiveness of specific rehabilitation methods in relation to I/I control and structural integrity, (2) the effectiveness of I/I control that might be expected on a systemwide basis, and (3) a procedure for measuring the effectiveness of systemwide structural integrity.

*S*ETTING THE OBJECTIVES OF SEWER REHABILITATION

For sewer rehabilitation to be successful, the conclusions of the system forensic engineering evaluation should be accurate. Determining the contribution of a rehabilitated element to the overall problems and the effect of rehabilitation on resolution of the overall problems are the most challenging and important elements in making a program effective.

INFILTRATION AND INFLOW CONTROL OBJECTIVES. The objective of the I/I analysis is to identify conveyance system segments where rehabilitation will return the highest investment. The achievement of this objective reflects the accuracy of the three basic I/I analysis steps: (1) quantifying I/I within subdrainage areas, (2) allocating the quantified I/I to the individual components within each subdrainage basin, and (3) estimating the effectiveness of rehabilitating specific elements in reducing overall I/I.

Selecting the Design Condition. Because I/I is linked to antecedent rainfall and specific storm characteristics, the rainfall or storm occurrence frequency chosen as a base for the analysis needs to be clearly defined. For example, I/I reduction values of a specific rehabilitation effort will vary, depending on the specific storm magnitude or antecedent rainfall condition selected as the basis for the analysis. It is the instantaneous or peak hour I/I value that is of interest.

The peak flow defines the required design condition and the conveyance and hydraulically dependent treatment system capacity capital costs. Average values are of interest only to the extent that increased volumes affect operational and maintenance costs. This appropriate "design condition" will vary with local system and permit conditions. For example, the same rehabilitation effort will appear more successful if postinfiltration flows are compared with prerehabilitation conditions resulting from a 10-year storm with antecedent rainfall-elevated groundwater, compared to a 5-year storm with low groundwater or a sudden thaw of accumulated snow on frozen versus unfrozen ground.

Allocating System Components for Quantified Infiltration and Inflow. A quantification of the various I/I values through the seasons provides additional evidence to assist in the accurate allocation of I/I to the system components. Examples of desired I/I values include minimum dry season, maximum month, and peak storm-influenced flows. The seasonal flow values and storm-influenced hydrograph characteristics with additional field data, including groundwater levels, smoke testing, and closed-circuit television

(CCTV) data, provide the basis for allocating the I/I to the system components.

Historically, the characteristics of the peak storm-influenced I/I flow regime would suggest evidence of a direct inflow source. Normally, it is actually leakage to service laterals that, by their shallow nature, show storm responsiveness characteristics similar to those from directly connected impervious areas. A sample infiltration allocation procedure is presented in Table 9.1.

Estimating the Effectiveness of Rehabilitation for Infiltration and Inflow Reduction. A common error in estimating the effectiveness of rehabilitation is to assume net systemwide effects will be equal to the sum of the I/I values initially allocated to specific rehabilitated components. Consideration has to be given to the "fluid" nature of the I/I sources, particularly if rehabilitation is limited to specific components in the total system.

Rehabilitation in one area can result in raising the groundwater level, increasing the leakage in shallower sewers, and creating new leakage in previously adequate sewers because of the increased hydraulic head. This is particularly true where rehabilitation efforts have been limited to agency-owned sewers while ignoring privately owned service laterals. Historically, peak flows represent a surcharge condition, in which rehabilitation efforts will not register any overall flow reduction until peak flows have been reduced below the capacity of the limiting conveyance segment of the surcharged section.

Understanding the effectiveness of the sewer rehabilitation I/I control program is essential to making the right decisions regarding rehabilitation versus increasing conveyance and treatment plant capacity. Recent successful I/I reduction efforts have recognized these interrelated factors by adopting I/I

Table 9.1 Example collection system component infiltration allocation (Brown and Caldwell, 1989).

Infiltration component	Sewer system component		
	Interceptor	Local sewer	Service lateral
Dry season minimum	a	a	
Wet season maximum month	b	b	b
Peak flow			c

[a] Allocate proportioned to inch-diameter-mile below water table.
[b] Allocate incremental increase proportional to inch-diameter-mile below water table.
[c] Allocate incremental increase less smoke test/dye-water test confirmed inflow sources 100% to the shallow service lateral assuming source is temporary perched groundwater and rainfall percolating through the ground.

reduction rehabilitation approaches that take a total basin-by-basin approach to overcome the interference effects of more piecemeal efforts.

STRUCTURAL REHABILITATION OBJECTIVES. The objective of the structural investigation is to locate vulnerable segments and define factors contributing to structural deterioration. Selecting the appropriate rehabilitation for a specific location requires a full understanding of the contributing factors, such as corrosion, differential external loadings, void creep, groundwater pressures, piping leakage, and hydraulic surcharging. Locating the most vulnerable segments and ascertaining their rate of deterioration will determine the success of structural rehabilitation. This is particularly important where hydrogen sulfide corrosion is the most significant factor contributing to structural deterioration. Structural rehabilitation should be undertaken as planned maintenance activities instead of as emergency repairs.

Construction-Phase Quality Assurance

To ensure rehabilitation quality, the same steps are taken as in any construction activity but are modified to reflect the limited access nature of sewer rehabilitation.

The specialist nature and often proprietary methods present a challenge best met by developing a detailed understanding of proposed materials and characteristics, step-by-step installation procedures, and field-specific installation conditions. This can help develop construction documents and inspection and testing procedures that ensure control over quality and cost.

The major elements of a successful construction quality assurance program are material quality control, existing sewer condition, safety, preparation of sewers, installation, annular grouting, payment, and construction inspection.

MATERIAL QUALITY CONTROL. Standards have been developed for the more common lining materials over the last few years, although standards will not normally be available in the case of newly emerging materials. The approach with the nonstandard material is to work from the manufacturer's standard specification and add clauses to provide an equivalent degree of assurance as reference to an adopted standard. Sample clauses should cover tests on the material type to prove its suitability and provide material properties for design calculations and tests on the product. This would prove that the particular size and shape of the lining can be manufactured with the properties already established in the material-type test.

EXISTING SEWER CONDITIONS. The unique nature of each rehabilitated section requires the engineer to provide in the contract documents all relevant information. This includes current structural and surface conditions, materials deposition, dimensional deviation, and sewer flows. Where bypass pumping is required, options and restrictions and conditions under which contracted operations may be temporarily terminated must be clearly defined.

SAFETY. Acceptable safety standards are essential to construction-phase quality assurance. Contracts should clearly define any special or unusual hazards and elements of the safety plan to be developed by the contractor. Though it is the contractor's legal responsibility to provide and maintain healthy and safe conditions, this does not fully protect the engineer and owner when there are accidents.

PREPARATION OF SEWERS. The degree of preparation should reflect the requirements of the particular renovation method. Prerenovation cleaning facilitates entry of linings and, by removing surface slimes, obstructions, and loose debris, promotes a bond between the existing sewer and the renovation material. This provides the opportunity to conduct preinstallation checks, particularly in segments where access or cross-sectional tolerances are limited.

INSTALLATION. Installation specifications are specific to the particular renovation method used, though key areas should be addressed. These include storage and handling of materials to preclude ultraviolet degradation, avoidance of concentrated offloading and stacking loads and surface damage, positioning of joints—particularly longitudinal for segmented systems—to ensure structural integrity, jointing techniques and environmental conditions to ensure quality, acceptable cutting methods, and installation tolerances including grade, lining thickness, and final cross section.

ANNULAR GROUTING. Many renovation techniques rely on annular grouting to provide overall structural integrity. Because there is no reliable, inexpensive method of ensuring that the grout has penetrated and filled all the annular voids, close control of the grouting operation and of grouting materials is required to ensure a high degree of confidence in the finished work. Grout material testing should encompass workability, density, bleed percentage, setting time, and compressive strength.

Grout installation specifications should encompass grout preparation, type of pumping equipment, including pressure range and capacity, and acceptable injection methodology. Typically, injection procedures include grout volume estimates, infiltration quantity and grout impact estimates, linear and circumferential grout sequencing, displaced air venting, grout pressure monitoring and recording inspection, grouting operation lining distortion inspection, and grout volume monitoring and recording requirements.

CONTRACTOR PAYMENT. Renovation work requires flexibility in bid documents to manage unforeseen circumstances when they arise. Adoption of a unit price bid schedule based on the engineer's judgment of specific quantities or units of each work item is the preferred contracting approach. Because of the imprecise nature of renovation projects, allowances in evaluating alternatives should reflect the degree of uncertainty of alternative renovation techniques. When a specific contract is awarded, a contingency budget of up to 50% of the initial bid may be required to cover deviations in final measured quantities.

CONSTRUCTION INSPECTION. Construction inspection on behalf of the owner by qualified inspectors is essential to the construction-phase quality assurance program. Characteristics of most sewer rehabilitation methods present a challenge for the construction inspector because many of the critical sewer rehabilitation steps take place in remote inaccessible locations and, even where access is available, often cannot be directly observed. In these situations, indirect inspection methods such as CCTV, air and hydraulic testing, and (where grouting is undertaken) close monitoring of grout material, grouting quantity and pressures plus postinstallation dimensional checks must be considered. The National Association of Sewer Service Companies has developed inspection guidelines that outline the key elements of an effective inspection program.

*M*EASURING THE EFFECTIVENESS OF INFILTRATION AND INFLOW CONTROL

To determine the effectiveness of rehabilitation I/I control, sewer system flow must be measured both before and after rehabilitation. Because of the difficulty of accurately measuring the constantly changing flows in a sewer system, the measurement of I/I is not exact. Methods for carrying out these measurements are described in Chapter 6.

The most common way to calculate the effectiveness of rehabilitation is to subtract flows after rehabilitation from flows before rehabilitation and to divide the result by flow before rehabilitation minus the dry weather flows. In addition to accurate flow measurements, such a calculation requires that either groundwater conditions and rainfall durations and intensities be similar for measurements of flow made before and after rehabilitation or that a dependable method is available for normalizing flows to a consistent set of conditions.

The long-term monitoring conducted at treatment plants and major pumping stations can provide overall system information for comparisons before and after rehabilitation, particularly for assessing long-term trends of I/I levels. These data typically are not appropriate for assessing the effectiveness of local I/I control efforts.

METHODS FOR NORMALIZING FLOW DATA. There are several methods for normalizing the impacts of rainfall to allow equivalent comparison for disparate weather years, including the following simulation technique that considers specific antecedent rainfall characteristics and can be used as a tool to regulate years of different precipitation and snowmelt patterns (Brown and Caldwell, 1988).

Infiltration Simulation. Water infiltrating a sewer system depends on the depth of groundwater above the sewer (a long-term continuous process) or water available in the saturated zone surrounding the sewer (a more short-term process). The rate of infiltration also depends on the physical condition of the sewer.

Empirical evidence suggests that infiltration depends on a time series of past rainfall, from a relatively immediate response in areas where the sewer is close to the surface and in poor condition, to a long-term response, where the groundwater table influences the process. In this simulation, the infiltration process is described by adding components caused by rainfall over the periods of the previous 24 hours, the previous 7 days, the previous 30 days, and the previous 6 months. The longer time components reflect the influence of changes in the groundwater table. The technique is equivalent to that used in the U.S. Environmental Protection Agency (U.S. EPA) Storm Water Management Model (SWMM). The technique is shown in the following equation:

$$\text{Infiltration} = C_d * R_1 + C_w * R_7 + C_m * R_{30} + C_s * R_{180} + \text{Basin}_f \quad (9.1)$$

Where

C_d	=	a coefficient describing the response to rainfall over the past 24 hours;
C_w	=	a coefficient describing the response to rainfall over the last 7 days;
C_m	=	a coefficient describing the response to rainfall over the last 30 days;
C_s	=	a coefficient describing the response to rainfall over the last 180 days;
$R_1, R_7, R_{30},$ and R_{180}	=	corresponding rainfall depths over the respective periods; and
Basin_f	=	minimum expected infiltration rate.

The above equation is then used to estimate the infiltration rate during any hour of the simulation. In practice, the various coefficients are found by calibration. This involves observing the response of infiltration to past rainfall by direct plots and multiple regression analyses after removal of direct inflow from the observed record. The various components are fine tuned by comparing the simulation with recorded flows over an extended period.

Inflow Simulation. The source of inflow is from a tributary service area directly connected to the sanitary sewers by illicit or purposeful connections. The connected area ordinarily is assumed to be only impervious surface in standard I/I analyses. Some pervious surface may be connected, as in combined systems or where cross connections to a stormwater system exist.

To simplify the simulations, inflow can be computed by the rational method. Inflow from impervious surfaces in any hour of the simulation is simply the product of the rainfall depth that occurred that hour times the impervious surface area. Where pervious surfaces are known to be connected, inflow from these sources is generated in the same way after subtraction of a rainfall abstraction to account for surface infiltration. Pervious area runoff is the product of the hourly rainfall less the estimated infiltration rate times the estimated pervious surface connected. Pervious surface runoff is zero if the rainfall does not exceed the estimated surface infiltration capacity. The infiltration capacity may be varied to reflect antecedent rainfall, depending on the sophistication of the analysis.

The simulation requires an estimate of the directly connected pervious and impervious surface area. Alternatively, these values may be determined by calibration to match observed hydrographs with simulated values, an approach typically adopted using this analytical approach.

Sanitary Flow Simulation. The domestic wastewater component of flow is simulated by applying an average diurnal normalized hydrograph to the average estimated domestic wastewater flow. The resulting hydrograph is normalized by division by the average value to give a hydrograph of percentage of average flow by hour of the day.

Storm Inflow Comparison. Good correlation has been shown to exist between the logarithm of rainfall intensity and logarithm of peak inflow, suggesting that groundwater is not a significant variable affecting inflow (Nogaj and Hollenbeck, 1981). This method provides a way of extrapolating inflow results before and after rehabilitation to a similar rainfall intensity or storm recurrence interval. This method also normalizes data from differing storms and requires approximately one rainy season of data. Other researchers used this method at two locations (Nelson and Bodner, 1981). Although the data show scatter, the results can be useful in measuring inflow control effectiveness.

FLOW-MONITORING CONSIDERATIONS. Flow monitoring for determining effectiveness must include a large enough part of the sewer system to ensure that sources eliminated in one location are not simply entering somewhere else by traveling through the bedding material surrounding the pipes to an area that has not been rehabilitated. For small systems, measurement of total flows at the treatment plant may provide meaningful results. For large systems, some segment of the total system should be selected.

STUDIES OF REHABILITATION EFFECTIVENESS FOR INFILTRATION AND INFLOW CONTROL. The most extensive survey ever conducted to determine specifically the effectiveness of sewer rehabilitation programs to reduce I/I was carried out under the support of U.S. EPA (Conklin and Lewis, 1980). The study was initiated as a result of evidence that I/I removal programs were not proving to be as effective as predicted and that the programs were more time consuming and expensive than originally expected. Most of these initial efforts reflected a lack of overall understanding of the nature of I/I and key factors, such as the contribution of service laterals to peak I/I and the movement of infiltration sources from rehabilitated to nonrehabilitated segments (Steketee, 1981). One community that was successful in achieving its I/I reduction goal undertook a systemwide approach to the rehabilitation effort, thus eliminating the issue of moving the problem to nonrehabilitated areas.

Information presently available indicates there are effective methods for reducing I/I in sewers, but problems and conditions that influence I/I control can exist in a sewer system. Service laterals, for which remedial measures recently have been developed, represent a significant part of the problem. Not only do laterals contribute their own I/I, but frequently their connection to the main sewer causes leakage, which most of the postrehabilitation remote reconnection procedures cannot remedy. In addition, any joints near laterals cannot be grouted internally by the commonly used inflatable-packer-based methods.

Another major factor in the poor overall effectiveness of piecemeal rehabilitation is the migration of groundwater infiltration sources along pipes within the granular bedding material. When this occurs, any partial rehabilitation is unlikely to be successful. Depending on grade, the water can move to unrehabilitated openings, causing leakage to continue (Steketee, 1981). There is a good chance that the net effectiveness of control of I/I contributing to the surcharged part of a system will be overestimated. No net reduction of I/I will be observed until I/I levels are reduced to the capacity of the surcharged segment. The migration problem is difficult to quantify. Because no dependable predictive means exist for this phenomenon, consideration must be given to the potential for migration in any rehabilitation program. Taking a total basinwide approach is a proven effective way to limit migration effects to the outer edges of the rehabilitated basin.

*M*EASURING THE EFFECTIVENESS OF REHABILITATION FOR STRUCTURAL INTEGRITY

As with measures to control I/I, the measurement of success of structural integrity-focused rehabilitation efforts includes the effectiveness of the specific repairs and, from a systemwide perspective, reduction and eventual elimination of emergency response repair activities.

SPECIFIC REPAIR EFFECTIVENESS. The initial measure of the effectiveness of structural rehabilitation is reinstatement of the conduit. Long-term effectiveness can be assessed by monitoring the recurrence of corrosion, deformation of the repaired segment, settlement of the line or the surface above, or leakage.

OVERALL SYSTEM INTEGRITY. The measure of success of structurally focused rehabilitation is when knowledge of the conveyance system condition allows for structural repairs within a planned schedule. This is best accomplished when evaluation of structural condition indicators is incorporated into maintenance activities of the sewer utility, as described at the end of this chapter.

*E*XPECTED EFFECTIVENESS OF SPECIAL SEWER REHABILITATION METHODS

Historically, the overall effectiveness of sewer rehabilitation has been compromised more by the incompleteness of rehabilitation efforts than failure of a specific method. In evaluating specific rehabilitation efforts, the ability to address connector points, service laterals, and groundwater migration potential must be considered in addition to addressing repair of the main pipe barrel.

There are several major rehabilitation methods, descriptions of which are given in Chapter 7.

SEWER REPLACEMENT. The most expensive method of sewer rehabilitation is replacement. Where serious structural damage has occurred or differential settlement has affected sewer grade, this may be the only reasonable approach. Effectiveness of control of infiltration in the replacement pipe is the same as for any other new sewer. Depending on the length of sewer replaced and the severity of water migration outside the new pipe to the parts of the sewer adjacent to the replaced section, infiltration removal will be less.

Replacement is also an option for rehabilitation of laterals. If no lateral replacement is proposed, the approach taken in reconnection of laterals can significantly reduce the net effectiveness. Extreme care must be used in making the reconnections and trench backfilling to avoid leakage between the main barrel and the laterals. Poor connections of the newly constructed sewer to service laterals can result in no net reduction of infiltration. To achieve overall effectiveness in parallel with the main barrel replacement, a program of lateral testing and rehabilitation should be carried out. Smoke and dye testing of all laterals before hookup will ensure existing inflow sources are not reconnected to the rehabilitated section and thus eliminated.

SEWER RELINING. Lining and Sliplining. Sewer relining involves placing a layer of piping material inside an existing pipe by inserting a slightly smaller pipe inside the existing pipe. Relining can be cheaper than replacement, excavation being limited to access for liner insertion and for nonperson entry projects for reconnecting laterals. In the case of sliplining, the cost difference depends on the number of laterals to be reconnected; the more laterals, the smaller the difference.

There is a wide variety of lining materials, ranging from cement applied directly to the inside of the existing pipe to modern plastic materials. Continuous plastic pipe linings can reduce infiltration to zero, though the net I/I control effectiveness of sliplining is a function of the integrity of sealing the annular space between the outside of the liner and the inside of the original pipe. Continuous grouting of the annular space will produce a more reliable seal than just packing the annular space at manhole pipe protrusions. The long-term integrity of high-density polyethylene has been shown; however, long-term net effectiveness will be more a function of the life of the annular space sealant.

Piping materials that are inserted but use the methods of joining pipe sections have a greater chance for leakage but still can be highly resistant to infiltration with effective annular space sealing and jointing technique. Where existing lateral to main line connections are sound, hookup of laterals is limited to cutting out the part of the lining covering the lateral and sealing the annular space. The integrity of this sealing step is a major factor in overall infiltration reduction effectiveness. If the existing lateral to main line connection is not sound, a new lateral connection directly to the liner by a pipe saddle arrangement can achieve the best results. Typically, this will require external exposure of the lateral, requiring extreme care in the backfilling operation. Lining and sealing the annular space and careful lateral reconnections can be as effective in controlling I/I as replacement methods.

With the exception of thin coatings of mortar, sewer lining methods have the capability of maintaining or improving the structural stability of the sewer system and often are applied partly for that purpose. There is a range of materials available to meet structural needs when sliplining. The effectiveness for attaining structural integrity is increased by completely grouting the annular space between the new pipe and the original pipe.

Inversion Lining. Because it has close contact with the inside of the original pipe, inversion lining eliminates annular space leakage. If the part of the lining covering laterals is cut out properly, leakage around the laterals can be reduced to a low value. Lack of care in this step can result in poor infiltration control. As with other lining methods, the cost of replacement is strongly affected by the degree of difficulty and the cost involved in excavating for pipe replacement. Inversion lining can be as effective in controlling I/I as replacement methods and does not require excavation to reconnect laterals if the existing lateral to main line hookup is in sound condition.

Inversion lining can be used for lining manholes and should exhibit the same high degree of infiltration reduction shown in sewer pipes. Openings to the sewers entering a manhole should be made carefully, as leakage could significantly reduce the overall effect of lining.

The strength of the inversion liner can be varied by changing the thickness or resin type, thus enabling it to overcome some structural instability. Inversion lining has been used to stabilize brick sewers where the mortar was badly eroded and bricks had become loose or were even missing.

SEWER SEALING. Chemical grout sealers for internal grouting of small to medium sewers are widely accepted in the sewer maintenance industry, with even relatively small utilities owning their own grout packers and sealing equipment. The effectiveness of chemical grouting to seal a leaking joint is a function of the condition and structural stability of the pipe, the surrounding backfill material, and the quality of workmanship. Chemical grouting using conventional packing equipment is most effective where the failed element is the joint, not the pipe material.

Where the grout is correctly applied, it is effective in preventing infiltration, approaching 100% for the specific sealed joint. However, this high degree of effectiveness only applies to the sealed joint, not necessarily to the section of pipe.

Leakage from service laterals, joints close to service laterals, adjacent pipe sections, and defects not correctable by the sealing procedure can render infiltration removal less effective. Though specific joints may have passed an air test, the sealing of adjacent joints can cause leakage of other adjacent joints. Improper assessment of these leakage sources has been a factor in the overestimation of infiltration control.

Assessment of chemical grouting effectiveness involves the same analytical considerations given to other rehabilitation methods that only address the main barrel of the pipe. Because of the joint-by-joint nature of the rehabilitation method, the overall effectiveness of a one-shot grouting effort will be less effective than replacement or lining methods. However, when grouting is incorporated to a regular maintenance program, it can control main pipe barrel leakage like a new pipe.

The long-term effectiveness of a particular sealed joint is influenced by factors such as groundwater changes, pipe surcharge conditions, the potential for pipe trench differential settlement, and the quality of the original grouting operation.

Sewer sealing does not improve the structural stability of the system and should not be used in sewers in poor structural condition. Unless the pipes are structurally stable, an effective sealing job probably could not be done and new leaks would develop rapidly.

External grouting methods of sealing joints leaking will be less effective in controlling infiltration than direct, internal application techniques.

SERVICE LATERAL REHABILITATION. Service laterals can constitute a serious source of both infiltration and inflow. They can contribute up to 75% or more of peak infiltration flows. The rehabilitation methods applied to the main sewer line, including sliplining, inversion lining, and grouting, have been adapted for rehabilitating service laterals in addition to excavation and replacement. Although these methods can provide a high degree of infiltration control within the pipe itself, there is no guarantee they will control deliberate sources of inflow and, in some cases, infiltration that occurs within the building before connection to the service lateral. Inflow control requires an effective disconnection and enforcement program.

In addition to I/I from the laterals, infiltration frequently results from a leaky connection of the lateral to the main sewer and leakage at main sewer joints close to the lateral; effective I/I control requires testing and repairing these sources of infiltration. Remedial methods have been developed to seal leaks at the lateral to main sewer hookup that conventional packing and grouting equipment could not seal. Such sealing technologies can be expected to be subject to the same limitations as conventional pipe sealing, requiring regular testing and regrouting to ensure long-term effectiveness.

INFLOW CONTROL. Inflow is controlled by disconnecting the pathway by which storm-generated surface waters enter the sewer. Typical pathways are manhole covers, catch basins, area drains, and roof drain downspouts. The actual peak flow reduction effectiveness of a specific inflow control method in terms of reduced peak flow rates will be a function of the particular storm characteristics and, to a lesser extent, to antecedent rainfall and ground freezing conditions (Milwaukee Metropolitan, 1981).

Manholes. Manhole covers containing vent and pickholes can be significant sources of inflow when they are located in the path of surface runoff. Replacement with a waterproof, gasketed cover is estimated to be 90% effective in reducing inflow.

Manholes frequently leak between the frame and corbel, especially if there is heaving of the pavement from freezing. Use of elastomeric sealants poured or troweled on the outside of the manhole or elastic sleeves is estimated to be 90% effective in reducing leakage. Application of an adhesive sealant to the interior of the corbel and joint beneath the flange of the manhole frame is estimated to be only 75% effective because water can still enter the space between the frame and corbel, increasing the chance for seal failure from frost action.

Catch Basins. Catch basins and area drains connected to sanitary sewers can contribute large amounts of inflow. Plugging the connection to the sanitary system and reconnection to a storm drain is estimated to be 90% effective in reducing inflow. The effectiveness is estimated to be less than 100% to compensate for migration of some water to other parts of the sanitary sewer system.

Roof Drains. Downspouts or roof drains are frequent sources of inflow. Disconnection of these from the sanitary system with reconnection to a storm sewer is estimated to be 90% effective in reducing inflow, with the remaining 10% finding its way to the sewer system by other routes. Where the disconnected downspout is discharged on the ground surface rather than being connected to a storm sewer, the inflow reduction is likely to be significantly less (possibly zero if service laterals serving the property are in poor condition).

Other. Sump pump and foundation drain connections to sanitary sewers represent other significant sources of inflow. Disconnection of these sources and reconnection to storm sewers was observed to result in approximately 75% inflow reduction. Any discharge of these disconnected sources to the ground surface prevents net reduction. To maintain long-term effective control requires an effective enforcement program to preclude reconnection.

OVERALL SEWER SYSTEM EFFECTIVENESS OF INFILTRATION AND INFLOW CONTROL. Because of a lack of understanding of the factors affecting I/I removal effectiveness, overestimation of the removal for total sewer systems or parts of systems has occurred. Because of the significant involvement of groundwater, particularly storm-influenced groundwater, the problem has been with overestimation of the effectiveness of infiltration control more than inflow.

The most certain approach for accurately predicting the overall effectiveness of a proposed infiltration control system is to conduct pilot studies of the effectiveness of the major rehabilitation methods. Results then can be extrapolated for the remainder of the system. This is being done with increasing frequency, especially in large cities. Without pilot data, the engineer must consider sources of leaks, such as service laterals, that have not always been adequately accounted for in the past.

In two specific cases of pilot studies of joint grouting—one in the Washington, D.C., area and one in Westchester County, New York—the range in infiltration related to service laterals and manholes (Bonk *et al.*, 1978, and Sewer System Evaluation, 1982). This wide difference indicates the variability that can be encountered in the field, the need to consider all possible infiltration sources, and the need to be conservative in estimating the effectiveness of infiltration control when actual data are not available on leakage rates from all significant sources.

There is an acknowledged need to eliminate all infiltration sources in sewers being rehabilitated. Although there are differences of opinion over the influence that migration can have over long distances, there is little doubt that for short distances it can cause significant problems. The realization of the need for completeness in infiltration control began with sewer sealing, but now must be extended to service laterals, manholes, and any other significant leakage sources. Where migration occurs over long distances, more control will be gained by thoroughly rehabilitating selected long segments of the system rather than many short segments. The latter alternative could produce poorer results for the same cost.

The problem of accurately estimating the effectiveness of inflow control over a sewer system generally is less difficult than estimating infiltration control. The exclusion of any inflow sources from sanitary sewers would seem to be 100% effective, but in many cases there is an opportunity for some of the water to reenter the sewer. Correction for reentry must be considered when estimating the effectiveness of remedial measures.

*C*ONTINUING SEWER MAINTENANCE

Sewer systems require constant maintenance; all structures wear out in time, and pipelines are no exception. Because these pipes are underground, signs of wear are not obvious until there is a failure. Failures start with cracking, lateral deflection, offset joints, deteriorated joint material and, in the presence of corrosive atmospheres, spalling of concrete and corrosion of reinforcing steel.

Typically, community residents only see the result of prolonged neglect, such as backed up sewers, local flooding, cracked pavement, and (in certain extreme cases) collapsed streets. Without adequate maintenance and vigilance on customer practices, levels of I/I will continue to increase and structural failures will occur more frequently. Recognizing that service laterals can contribute up to 75% or more of peak I/I, rigid quality control and testing requirements should be placed on their initial construction. Sewer ordinances should require clean-out access at the building edge and property line to ease periodic testing and should clearly identify the property owner's responsibility for service lateral repair as required to meet specific air test standards. New local sewers should be inspected by CCTV and air tested and service laterals should be air tested before they are accepted by the local sewerage agency.

Whether a system is new or has recently been rehabilitated, continuing maintenance is necessary to ensure structural integrity and to prevent gradual increases in I/I. Many problems undoubtedly could be taken care of less expensively if they were addressed as part of planned maintenance instead of allowed to deteriorate to critical levels. Budgeting for sewer system maintenance typically takes a lower priority than capital funding for a wastewater treatment plant or other more visible community problems. With the decline in outside federal and state funding support for wastewater rehabilitation and construction projects, it is now more important that municipalities and sewerage agencies fund preventive maintenance programs.

The lack of either communication between field crews and agency management or a forum for bringing issues to elected officials often is a barrier to adequate funding. Following are some approaches for identifying the appropriate level of budgeting for preventive maintenance, incorporating condition assessment and response into sewer maintenance cleaning activities, and using annual reporting to inform elected officials.

DATA NEEDS FOR BUDGETING PREVENTIVE MAINTENANCE.

To ensure that maintenance activities of a facility are adequately budgeted for, it is necessary to establish an effective method of predicting the long-term sewerage system rehabilitation and funding needs to maintain the structural and hydraulic integrity of the system. Factors involved in any prediction approach include determination of the useful life of individual system components and an inventory of the system components, including condition assessment, corrosion assessment, and determination of critical and noncritical segments.

A study undertaken by the city of Seattle evaluated a number of approaches for forecasting its rehabilitation budgets and programs (Brown and Caldwell, 1987). The city used statistical analysis of physical assessment records for critical and noncritical component life expectancies. Using actuarial techniques, survival curves were developed for individual sewerage system elements. These curves were developed using both "annual rate" and "individual unit" methods. Both methods require an inventory of the existing system by component or unit, the number of units replaced or rehabilitated each year with their age, and the number of units in service and their respective ages.

Based on the city analysis, the average life of the sewer system was estimated at approximately 200 years, indicating annual budget needs of approximately 0.5% of the replacement value for sewerage rehabilitation. This relatively high average life value reflects local factors of minimum corrosion, the quality of initial construction, and good maintenance practices.

Regardless of the assessment techniques used, they all require that a complete and easily accessible sewerage system component inventory be established. There are a number of proprietary software packages for sewerage system information management that include collection system inventory, maintenance history, structural condition indicators, flow records, and maintenance scheduling.

SEWER MAINTENANCE ACTIVITIES. The conventional focus of sewer maintenance crews is on flushing and cleaning activities. With more widespread use of CCTV, many utilities now can inspect their systems, maintain condition records, assess effectiveness of corrosion control measures, and undertake air testing, grouting, and other leak control repairs. Using the day-to-day experience of the sewer maintenance field crews in a managed, focused way should yield a better understanding of the conveyance system than would an intermittent evaluation undertaken by outside staff.

ANNUAL REPORTING. As a collection system ages, levels of infiltration and inflow can be expected to increase gradually unless preventive measures are incorporated to regular maintenance activities. An effective management technique to track the trend of infiltration and inflow is through the utility annual report. For certain wastewater utilities, annual I/I reporting is a requirement of the National Pollutant Discharge Elimination System permit for operating the wastewater treatment plant. By prescribing peak rate, seasonal, and annual I/I in terms of sewered area, length of sewer, percent of treated flow, and number of customers, the long-term trends and effectiveness of rehabilitation efforts and consequences of continuous deterioration can be better understood by the utility's decision makers.

REFERENCES

Bonk, M.P., *et al.* (1978) Test Program for the Anacostia Sewerage System Rehabilitation. Rep. 1, Wash. Suburban Sanit. Comm., Hyattsville, Md.

Brown and Caldwell (1987) City of Seattle, Long Range Sewer Plan Sewerage Rehabilitation Study. Seattle, Wash.

Brown and Caldwell (1988) Municipality of Metropolitan Seattle, Carkeek Transfer/CSO Program I/I Analysis. Seattle, Wash.

Brown and Caldwell (1989) Municipality of Metropolitan Seattle, Local System I/I Evaluation, Phase II Report. Seattle, Wash.

Conklin, G.F., and Lewis, P.W. (1980) Evaluation of Infiltration/Inflow Program. Project 68-01-4913, U.S. EPA, Washington, D.C.

Milwaukee Metropolitan Sewage District (1981) Milwaukee Metropolitan Sewage District Sewer System Evaluation Survey. Milwaukee, Wis.

Nelson, R.E., and Bodner, R.L. (1981) Measuring Effectiveness of Infiltration/Inflow Removal. Paper presented at Am. Public Works Congress, Atlanta, Ga.

Nogaj, R.J., and Hollenbeck, A.J. (1981) One Technique for Estimating Inflows with Surcharge Conditions. *J. Water Pollut. Control Fed.*, **53**, 491.

Sewer System Evaluation Survey, Mamaroneck Sewer District (1982) Dep. of Environ. Facilities, Westchester County, N.Y.

Steketee, C.H. (1981) Demonstration of Service Lateral Testing and Rehabilitation Techniques Prepared for the U.S. EPA. EPA-600/2-85-131, U.S. EPA, Washington, D.C.

Appendix A
General Formulae

The designer must determine loading conditions when using various rehabilitation methods. Several formulae are provided in this section for designing performance parameters; for example, suggested push–pull lengths for sliplining, bending angles for handling and construction, hydrostatic collapse, and buckling resistance, along with recommended safety factors.

PIPE INSTALLATION (FUSED). The fused polyethylene (PE) pipe can be pulled by a cable attached to a pulling head fastened to the pipe. This prevents damage to the PE pipe. The length of the fused pipe that can be pulled will vary depending on field conditions and ease of access to the area. In general, the maximum pulling length for pipes with diameters of 300 mm (12 in.) and smaller is limited to 305 m (1 000 ft) and for larger pipe approximately 150 m (500 ft). Information about various pulling forces and lengths is desirable for design and estimating purposes. The maximum force that can be applied to a pipe on level ground can be determined by the following formula:

$$P_t = (\sigma_{max})A \tag{A.1}$$

Where

P_t = maximum pulling force, N·m (lb–ft);
σ_{max} = maximum allowable tensile stress, 6 900 kPa (1 000 psi); and
A = cross-sectional area of pipe, mm^2 (sq in.).

(Note: high-density polyethylene [HDPE] properties are specified by cell classification according to ASTM D-3350. Tensile strength normally is cell class 4, which is 20 700 to 24 000 kPa [3 000 to 3 500 psi]. The allowable tensile stress of 6 900 kPa [1 000 psi] provides a design factor of approximately 3 to compensate for installation stresses.)

The following formula can be used to determine the pulling length:

$$L = \frac{P_t}{f\alpha \, (SF)} \tag{A.2}$$

Where

- P_t = maximum pulling force, N·m (lb–ft);
- L = pulling length, m (ft);
- f = friction coefficient, 0.5;
- α = pipe weight, kg/m (lb/ft); and
- SF = 2.0 (for installation).

LONGITUDINAL BENDING (FUSED). Bending induced during the insertion step, in transporting pipe lengths from assembly sites to job sites, or permanent bends to accommodate line or grade changes should be limited to radii equivalent to a longitudinal strain recommended by the pipe manufacturer. The minimum allowable radius of curvature for any size and weight of pipe can be closely approximated from the following equation:

$$R_c = \frac{D_o}{2(\varepsilon_a)} \tag{A.3}$$

Where

- R_c = radius of curvature, mm (in.);
- D_o = outside diameter of inserted pipe, mm (in.); and
- ε_a = allowable axial strain, 1.5%.

This equates to

$$R_c = \frac{D_o}{2(0.015)} = \frac{D_o}{0.03} = 33 \, D_o$$

Note: the allowable axial strain has a design factor of at least 3; however, the manufacturer should be consulted regarding the recommended long-term allowable strain.

PIPE INSTALLATION (GASKETED). The gasketed pipe joint segments can be pushed or pulled into the existing pipeline from an insertion pit. The pipe joints should be inserted with the spigot end first and the bell end trailing. The push–pull bearing plate should be applied against the flat surface of the bell step to avoid damaging the bell. The maximum pushing or pulling length is determined by the longitudinal compressive strength of the pipe, which varies with the type of material and its design.

The access pit should be approximately 1.5 to 3 m (5 to 10 ft) longer than the standard 6-m (20-ft) pipe segment length. The width of the pit should be

0.6 to 1.2 m (2 to 4 ft) wider than the diameter of the existing pipe. In general, the maximum push–pull lengths for 450 mm (18 in.) and larger slipliner pipe is limited to 305 m (1 000 ft) in a dry sewer and about twice that in an active flowing sewer.

Existing pipeline conditions such as alignment and grade change and structural and corrosion conditions must be determined before the installation. The maximum push–pull force can be determined by the following formula:

$$P_c = \sigma_{max} A \tag{A.4}$$

Where

P_c	=	maximum push–pull force, N·m (lb–ft);
σ_{max}	=	maximum allowable compressive stress, kPa (psi); and
A	=	cross-sectional area of pipe, mm^2 ([sq in.] located at minimum cross section).

The following formula can be used to determine the push–pull length:

$$L = \frac{P_c}{f\alpha \, (SF)} \tag{A.5}$$

Where

P_c	=	maximum push–pull force, N·m (lb–ft);
L	=	estimated push–pull length, m (ft);
f	=	friction coefficient (0.5 in dry conditions, 0.25 in wet conditions);
α	=	pipe weight, kg/m (lb/ft); and
SF	=	2.0 (for installation).

HYDROSTATIC LOADS. When there is a possibility of groundwater above the pipe, the level of groundwater and its duration should be estimated and pipe of sufficient wall thickness to withstand the pressure without collapsing should be used. An appropriate safety factor of 1.5 should be used. The following equation can determine the needed wall thickness:

$$P = \frac{24 \, E_a I}{(1-\mu^2) d^3 (FS)} \tag{A.6}$$

Where

P	=	water head pressure, kPa (psi);
E_a	=	apparent (time-corrected) modulus, kPa (psi);
I	=	moment of inertia, mm^4/mm (in.^{4}/in.);
d	=	mean diameter, mm (in.);
μ	=	Poisson ratio; and
FS	=	safety factor, normally 1.5.

To determine the needed wall thickness:

$$P = \frac{2E_a (t)^3}{(1-\mu^2)(d)^3(FS)} \tag{A.7}$$

$$t = \frac{[P(1-\mu^2)d^3(FS)]^{1/3}}{[2 E_a]^{1/3}} \tag{A.8}$$

Where

t = wall thickness, mm (in.).

Formulae A.6 through A.8 apply primarily to solid wall pipe, such as high-density polyethylene, fiber-glass-reinforced plastic, reinforced plastic mortar, polyvinyl chloride, steel pipe, ductile iron pipe, and cured-in-place pipe.

The following mathematical modification of Formula A.7 may be used when using the dimension ratio (DR):

$$P = \frac{2 E_a}{(1-\mu^2)(DR-1)^3(FS)} \tag{A.9}$$

To determine the needed DR:

$$(DR-1)^3 = \frac{2E}{(1-\mu^2)P(FS)} \tag{A.10}$$

$$DR = \frac{[2E_a]^{1/3}}{(1-\mu^2)P(FS)} + 1$$

Note 1: apply an appropriate safety factor, normally 1.5.

Note 2: the manufacturer's long-term modulus values must be obtained through acceptable long-term testing.

RESTRAINED HYDROSTATIC LOADS. When the annulus between the slipliner pipe and the existing pipe to be lined exceeds 25 mm (1 in.), this space should be grouted to provide restraining support for the slipliner pipe. This support enhances buckling resistance by at least six times; hydrostatic buckling can be determined by using the following adjusted formulae:

$$P = \frac{2 K E_a t^3}{(1-\mu^2)d^3(FS)} \tag{A.11}$$

$$t = \frac{[P(1-\mu^2)d^3]^{1/3}}{[2KE_a]^{1/3}} \quad (A.12)$$

$$P = \frac{2KE_a}{(1-\mu^2)(DR-1)^3} \quad (A.13)$$

$$DR-1 = \frac{[2KE_a]^{1/3}}{[(1-\mu^2)PFS]^{1/3}} \quad (A.14)$$

Where

K = grout support factor.

Grout the annular space between the outside diameter of the installed liner pipe and the inside diameter of the existing pipe with a cement- or chemical-based grout. During grout placement, ensure that the safe grouting pressure is not greater than one-fourth the pipe stiffness value. The pipe stiffness is determined in accordance with ASTM D-2412.

STRUCTURAL LOADINGS. When the original pipeline is considered fully deteriorated, it is necessary to determine the buckling loadings. The AWWA C-950 buckling formula is as follows:

$$q_a = \frac{C}{N} \frac{[32 R_w B' E' E_L I]^{1/2}}{D^3} \quad (A.15)$$

Where

q_a = allowable external pressure on pipe, kPa (psi); and
C = ovality reduction factor, as follows:

Pipe deflection, %	C
1	0.91
2	0.84
3	0.76
4	0.70
5	0.64
6	0.59
7	0.54
8	0.49
9	0.45
10	0.41

N = safety factor (normally 2.5);
R_w = water buoyancy factor $[1 - 0.33(H_w + H)]$;

General Formulae

H_w = height of groundwater above pipe, m (ft);
H = height of soil above pipe, m (ft);
B' = soil support factor, kg (lb), $\dfrac{1}{1 + 4e^{-0.065H}}$;
E' = soil modulus, kPa (psi);
E_L = flexural modulus, long-term;
I = moment of inertia, mm^4/mm (in.^{4}/in.); and
D = mean pipe diameter, mm (in.).

For normal pipe loadings, the following is provided:

$$q_t = \gamma_w H_w + \frac{R_w W_s}{D} + \frac{W_L}{D} \leq q_a \qquad (A.16)$$

Where

q_t = total external pressure on pipe, kPa (psi);
γ_w = water density, kg/mL (lb/cu in.);
H_w = height of groundwater above pipe, mm (in.);
R_w = water buoyancy factor;
W_s = soil load on pipe (kg/mm [lb/in.], $\gamma_s D H_s$);
γ_s = soil density, kg/mL (lb/cu in.);
D = mean pipe diameter, mm (in.);
H_s = height of soil above pipe, mm (in.);
W_L = live load on pipe, kg/mm (lb/in.); and
q_a = allowable external pressure on pipe, kPa (psi).

Note: for solid wall pipe sections, $I = t^3/12$. For profile wall sections, it will be necessary to calculate the moment of inertia using the appropriate formulae.

Appropriate values should be determined by the designer in the calculations.

Appendix B
Glossary

abrasion Wear or scour by hydraulic traffic.

abrasion and scratch resistance Ability of a material to resist damage in the form of scratches, grooves, and other minor imperfections.

abutment A wall supporting the end of a bridge or span, sustaining the pressure of the abutting earth.

acceptance Concurrence by the owner that the work is complete in accordance with the contract documents.

addenda (1) Written or graphic instruments issued before the execution of the agreement that modify or interpret the contract documents. (2) Drawings and specifications, by addition, deletions, clarifications, or corrections.

additive A substance added in a small amount, usually to a fluid, for a special purpose, such as to reduce friction or corrosion.

aerial sewer An unburied sewer (generally a sanitary sewer), supported on pedestals or bents to provide a suitable grade line.

aerobic Presence of unreacted or free oxygen (O_2).

aggressive A property of water that favors the corrosion of its conveying structure.

aggressive index (AI) A corrosion index established by the American Water Works Association Standard C-400 as a criterion for determining the corrosive tendency of the water relative to asbestos-cement pipe. It is calculated from the pH, calcium hardness (H), and total alkalinity (A) by the formula $AI = pH + \log(AH)$.

agreement The written agreement between the owner and the contractor covering the work to be performed. The contract documents are attached to and made a part of the agreement. Also designated as the contract.

alkalinity The capacity of water to neutralize acids; a measure of the buffer capacity of water. The major portion of alkalinity in natural waters is caused by hydroxide, carbonates, and bicarbonates.

ASTM A scientific and technical organization formed for the development of standards on the characteristics and performance of materials, products, systems, and services and the promotion of related knowledge.

anaerobic An absence of unreacted or free oxygen [oxygen as H_2O Na_2SO_4 (reacted) is not free].

angle of repose The angle that the sloping face of a bank of loose earth or gravel or other material makes with the horizontal.

anode The opposite of cathode; the electrode at which oxidation or corrosion occurs.

apparent tensile strength A value of tensile strength used for comparative purpose that is determined by tensile testing pipe rings in accordance with ASTM D-2290. This differs from true tensile strength of the material due to a bending moment induced by the change in contour of the ring as it is tested. Apparent tensile strength may be at yield, rupture, or both.

apparent tensile yield The apparent tensile strength calculated for the yield condition.

application for payments The form furnished by the engineer for the contractor to request progress payments and an affidavit of the contractor that progress payments received from the owner on account of the work have been applied by the contractor to discharge, in full, all of the contractor's obligations stated in prior applications for payment.

approval Acceptance as satisfactory.

aqueous Pertaining to water; an aqueous solution is a water solution.

areaway A paved surface, serving as an entry area to a basement or subsurface portion of a building, that is provided with some form of drainage that may be connected to a sewer line.

available water Water necessary for the performance of work, which may be taken from the fire hydrant nearest the worksite depending on conditions of traffic and terrain that are compatible with the use of the hydrant for performance of work.

backfill density Percentage of compaction for pipe backfill (required or expected).

base (course) A layer of specified or selected material of planned thickness constructed on the subgrade (natural foundation) or subbase for distributing load or providing drainage or upon which a wearing surface or a drainage structure is placed.

base resin Plastic material before compounding with other additives or pigments.

batter The slope or inclination from a vertical plane, such as the face or back of a wall.

bedding The earth or other material on which a pipe or conduit is supported.

berm The space between the toe of a slope and excavation made for intercepting ditches or borrow pits.

bid The offer or process of the bidder submitted on the prescribed form setting forth the prices for the work to be performed.

bidder Any person, firm, or corporation submitting a bid for the work.

biological corrosion Corrosion that results from a reaction between the pipe material and organisms such as bacteria, algae, and fungi.

bituminous (coating) Of or containing bitumen such as asphalt or tar.

bonds Bid, performance and payment bonds, and other instruments of security furnished by the contractor and surety in accordance with the contract documents and in accordance with the law where the project is located.

boring An earth-drilling process used for installing conduits or pipelines or for obtaining soil samples for evaluation and testing.

bridge (1) A structure for carrying traffic over a stream or gulley or other traffic way, including the pavement directly on the floor of the structure. (2) A structure measuring 3 m (10 ft) or more in clear span.

bridge plank (deck or flooring) A corrugated steel subfloor on a bridge to support a wearing surface.

brittleness temperature Temperature at which 50% of the tested specimens will fail when subjected to an impact blow.

building sewer The conduit that connects building wastewater sources to the public or street sewer, including lines serving homes, public buildings, commercial establishments, and industry structures. For example, (1) the section between the building line and the property line, frequently specified and supervised by plumbing or housing officials, and (2) the section between the property line and the street sewer, including the connection between the two, frequently specified and supervised by sewer, public works, or engineering officials. Referred to also as house sewer, building connection, service connection, lateral connection.

buoyancy The power of supporting a floating body, including the tendency to float an empty pipe, by exterior hydraulic pressure.

burst strength The internal pressure required to cause a pipe or fitting to fail within a specified time period.

butt fusion A method of joining polyethylene pipe in which two pipe ends are heated to a molten state and rapidly brought together under pressure to form a homogeneous bond.

bypass (1) An arrangement of pipes and valves whereby flow may be directed around a hydraulic structure or appurtenance. (2) A temporary setup to route flow around a part of a sewer system.

bypass pumping The transportation of wastewater flows around specific sewer pipe sections by any conduit to control wastewater flows in the specified sections without flowing or discharging onto public or private property.

caisson A watertight box or cylinder used in excavations for foundations or tunnel pits to hold out water so concreting or other construction can be performed.

camber Rise or crown of the center of a bridge, or flowline through a culvert, above a straight line through its ends.

cantilever The part of a structure that extends beyond its support.

carbon black A black pigment produced by the incomplete burning of natural gas or oil that possesses excellent ultraviolet protective properties.

catastrophic rainfall event Rainfall event of return frequency in excess of any sewerage design performance criteria, typically a 20- to 200-year storm.

cathode The electrode of an electrolytic cell at which reduction is the principal reaction. (Electrons flow toward the cathode in the external circuit.) Typical cathodic processes are cations taking up electrons and being discharged, oxygen being reduced, and the reduction of an element or group of elements from a higher to a lower valence state.

cathodic corrosion An unusual condition (especially with aluminum, zinc, and lead) in which corrosion is accelerated at the cathode because cathodic reaction creates an alkaline condition corrosive to certain metals.

cathodic protection (1) Preventing corrosion of a pipeline by using special cathodes (and anodes) to circumvent corrosive damage by electric current. (2) A function of zinc coatings on iron and steel drainage products; galvanic action.

cation A positively charged ion that migrates through the electrolyte toward the cathode under the influence of a potential gradient.

cavitation Formulation and sudden collapse of vapor bubbles in a liquid, usually resulting from local low pressures such as on the trailing edge of a propeller. This develops momentary high local pressure, which can mechanically destroy a portion of a surface on which the bubbles collapse.

cell Electrochemical system consisting of an anode and a cathode immersed in an electrolyte. The anode and cathode may be separate metals or dissimilar areas on the same metal. The cell includes the external circuit that permits the flow of electrons from the anode toward the cathode.

cellar drain A pipe or series of pipes that collects wastewater that leaks, seeps, or flows to subgrade parts of structures and discharges it to a building sewer or disposes it to sanitary, combined, or storm sewers. Also referred to as basement drain.

cell classification As specified by ASTM D-3350, a method of identifying plastic materials, such as polyethylene, where the cell classification is based on these six properties:
1. Density of base resin,
2. Melt index,
3. Flexural modulus,
4. Tensile strength at yield,
5. ESCR, and
6. Hydrostatic design basis and color.

Also, for polyvinyl chloride, as specified by ASTM D-1784 or D-4396, where the cell classification is based on these six properties:
1. Base resin,
2. Impact strength,
3. Tensile strength,
4. Modulus of elasticity in tension,
5. Deflection temperature under load, and
6. Chemical resistance.

change order A written order to the contractor authorizing an addition, deletion, or revision in the work within the general scope of the agreement or authorizing an adjustment in the agreement price or agreement time.

chemical resistance Ability to render service in the transport of a specific chemical for a useful period of time at a specific concentration and temperature.

chimney The cylindrical, variable height portion of the manhole structure having a diameter as required for the manhole frame. The chimney extends from the top of the corbel or cone to the base of the manhole frame and is used for adjusting the finished grade of the manhole frame.

circumferential coefficient of expansion and contraction The fractional change in circumference of a material for a unit change in temperature. Expressed as inches of expansion or contraction per inch of original circumference per °F.

coefficient of thermal expansion and contraction The fractional change in length of a material for a unit change in temperature.

cofferdam A barrier built in the water to form an enclosure from which the water is pumped to permit free access to the area within.

cohesive soil Soil that, when unconfined, has considerable strength when air dried and significant cohesion when submerged.

cold bend To force the pipe into a curvature without damage using no special tools, equipment, or elevated temperatures.

collector sewer A sewer located in the public way that collects wastewater discharged through building sewers and conducts such flows to larger interceptor sewers and pumping and treatment works. Also referred to as street sewer.

combined sewer A sewer intended as both a sanitary sewer and a storm sewer or as both an industrial sewer and a storm sewer.

compaction The densification of a soil by means of mechanical manipulation.

compounding The process through which additives and pigment are homogeneously mixed with the base resin in a separate and additional process to produce a uniform compound material for pipe extrusion.

compression gasket A device that can be made of several materials in a variety of cross sections and secures a tight seal between two pipe sections, for example, "O"-rings.

conductivity A measure of the ability of a solution to carry an electrical current. Conductivity varies both with the number and type of ions the solution carries.

conduit A pipe or other opening, buried or aboveground, for conveying hydraulic traffic, pipelines, cables, or other utilities.

consolidation The gradual reduction in the volume of a soil mass resulting from an increase in compaction.

contract documents The agreement, addenda, instructions to bidders, contractor's bid, the bonds, the notice of award, the general conditions, the supplementary conditions, special conditions, technical conditions, the specifications, drawings and modifications, and notice to proceed.

contracting officer The owner (grantee) or individual authorized to sign the contract documents on behalf of the owner's governing body.

contractor The person, firm, or corporation with whom the owner has executed the agreement.

contract price The total monies payable to the contractor under the contract documents.

contract time The number of calendar days stated in the agreement for completion of the work.

corbel or **cone** That portion of a manhole structure which slopes upward and inward from the barrel of the manhole to the required chimney or frame diameter. **Corbel** refers to a section built of brick or block; **cone** refers to a precast section.

core area That part of a sewer network containing the critical sewers and other sewers where hydraulic problems are likely to be most severe and require detailed definition within a flow simulation model.

corrosion The destruction of a material or its properties because of a reaction with its (environment) surroundings.

corrosion fatigue Fatigue-type cracking of metal caused by repeated or fluctuating stresses in a corrosive environment characterized by shorter life than would be encountered as a result of either the repeated or fluctuating stress alone or the corrosive environment alone; the combined action of corrosion and fatigue (cycling stress) in causing metal fracture.

corrosion index Measurement of the corrosivity of a water (such as the Langelier, Ryznar, or Aggressive Index).

corrosion rate The speed (usually an average) with which corrosion progresses; often expressed as though it were linear, in units of mdd (milligrams per square decimeter per day) for weight change, or mpy (mils per year) for thickness changes.

corrosion resistance Ability of a material to withstand corrosion in a given corrosion system.

cracks Crack lines visible along the length or circumference.

creep The dimensional change, with time, of a material under continuously applied stress after the initial elastic deformation.

crevice corrosion Localized corrosion resulting from the formation of a concentration cell in a crevice formed between a metal and a nonmetal or between two metal surfaces.

crew The number of persons required for the performance of work at a site as determined by the contractor in response to task difficulty and safety considerations at the time or location of the work.

critical sewers Sewers that would suffer the most significant consequences in the event of structural failure.

cured-in-place pipe (CIPP) A pipe containing an absorbing fabric surrounded by a cured thermoset resin formed within an existing conduit taking the shape of the existing pipe.

curing The conversion of liquid resin to a solid, usually by heat.

day A period of 24 hours measured from midnight to midnight.

deadman Buried anchorage for a guy or cable.

debris Soil, rocks, sand, grease, or roots in a sewer line, excluding items mechanically attached to the line such as protruding service connections, protruding pipe, and joint materials.

deflection Change in shape or decrease in vertical diameter of a conduit, produced without fracture of the material.

degradation A damaging change in the chemical structure of the plastic.

design life The length of time for which it is economically sound to require a structure to serve without major repairs.

design pressure The pressure calculated using a stress value equal to the long-term hydrostatic design basis times the appropriated service design factor. This is the maximum operating pressure the main should be subjected to during any operating cycle.

design stress rating The hoop stress used as a design basis for long-term service. The design stress rating is found by multiplying the hydrostatic design basis by the design service factor (0.5).

diaphragm A metal collar at right angles to a drain pipe for the purpose of retarding seepage or the burrowing of rodents.

dimple A term used in tight-fitting pipeline reconstruction, where the new plastic pipe forms an external departure or a point of expansion slightly beyond the underlying pipe wall where unsupported at side connections. The dimples are used for location and reinstatement of lateral sewer service.

discharge (Q) Flow from a culvert, sewer, or channel.

distribution lines Those facilities used to carry water from the transmission lines to the service lines, including water mains, distribution reservoirs, elevated storage tanks, booster stations, and valves.

ditch check Barrier placed in a ditch to decrease the slope of the flowline and, thereby, decrease the velocity of the water.

drainage Interception and removal of groundwater or surface water by artificial or natural means.

drawings The drawings that show the character and scope of the work to be performed and that have been prepared or approved by the engineer and are referred to in the contract documents.

ductility The extent to which a solid material can be drawn into a thinner cross section.

easement A liberty, privilege, or advantage without profit that the owner of one parcel of land may have in the hand of another. In this agreement, all land, other than public streets, in which the owner has sewer system lines or installations and right of access to such lines or installations.

easement access Areas within an easement to which access is required for performance of work.

effluent Outflow or discharge from a sewer or wastewater treatment facility.

elastic modulus A measure of the stress buildup associated with a given strain.

elongation The increase in length of a material stressed in tension.

embankment (or fill) A bank of earth, rock, or other material constructed above the natural ground surface.

embrittlement Loss of ductility of a material, resulting from a chemical or physical change.

emergency repair A repair that must be made while the main is pressurized or flowing.

end section Flared attachment on inlet and outlet of a culvert to prevent erosion of the roadbed, improve hydraulic efficiency, and improve appearance.

endurance limit The maximum stress that a material can withstand for an infinitely large number of fatigue cycles (see **fatigue strength**).

energy gradient Slope of a line joining the elevations of the energy head of a stream.

energy head The elevation of the hydraulic gradient at any section plus the velocity head.

engineer The person, firm, or corporation named as such in the contract documents; the "Engineer of Record."

environment The surroundings or conditions (physical, chemical, mechanical) in which a material exists.

environmental stress cracking The susceptibility to crack under the influence of specific chemical or mechanical stress.

epoxy Resin formed by the reaction of bisphenol and epichlorohydrin.

equalizer A culvert placed where there is no channel but where it is desirable to have standing water at equal elevations on both sides of a fill.

erosion Deterioration of a surface by the abrasive action of moving fluids. This is accelerated by the presence of solid particles or gas bubbles in suspension. When deterioration is further increased by corrosion, the term "corrosion-erosion" often is used.

erosion corrosion A corrosion reaction accelerated by the relative movement of the corrosive fluid and the metal surface.

exfiltration The leakage or discharge of flows being carried by sewers out into the ground through leaks in pipes, joints, manholes, or other sewer system structures; the reverse of **infiltration**.

existing linear feet The total length of existing sewer pipe in place within designated sewer systems as measured from center of manhole to center of manhole from maps or in the field.

extra high-molecular-weight high density (EHMWHD) As originally noted in ASTM D 1248, Grade P34 materials were specifically EHMW high-density polyethylene materials.

fatigue The phenomenon leading to fracture under repeated or fluctuating stresses having a maximum value less than the tensile strength of the material.

fatigue strength The stress to which a material can be subjected for a specified number of fatigue cycles.

field orders A written order issued by the engineer that clarifies or interprets the contract documents or orders minor changes in the work according to the terms of the contract.

filter Granular material placed around a subdrain pipe to facilitate drainage and at the same time strain or prevent the admission of silt or sediment.

flash point The temperature at which a material begins to vaporize.

flexible Readily bent or deformed without permanent damage.

flexural modulus The slope of the curve defined by flexural load versus resultant strain. A high flexural modulus indicates a stiffer material.

flexural strength The strength of a material in bending expressed as the tensile stress of the outermost fibers at the instant of failure.

flow attenuation The process of reducing the peak flow rate in a sewer system by redistributing the same volume of flow over a longer period of time.

flow control A method whereby normal sewer flows or a portion of normal sewer flows are blocked, retarded, or diverted (bypassed) within certain areas of the sewer collection system.

flow reduction The process of decreasing flows to a sewer system or of removing a proportion of the flow already in a sewer system.

flow simulation The modeling of flows in surface water or combined sewer systems using a dynamic digital model.

fold and form pipe A pipe rehabilitation method where a plastic pipe manufactured in a folded shape of reduced cross-sectional area is pulled into an existing conduit and subsequently expanded with pressure and heat. The reformed plastic pipe fits snugly to and takes the shape of the inside diameter (ID) of the host pipe.

fouling An accumulation of deposits. This term includes accumulation and growth of marine organisms on a submerged metal surface and also includes the accumulation of deposits (usually inorganic) on heat exchanger tubing.

foundation drain A pipe or series of pipes that collect groundwater from the foundation or footing of structures and discharge it to sanitary, storm, or combined sewers or to other points of disposal for the purpose of draining unwanted waters away from such structures.

fracture mechanics A quantitative analysis for evaluating structural reliability in terms of applied stress, crack length, and specimen geometry.

fractures Cracks visibly open along the length or circumference of the conduit with the pieces still in place.

galvanic cell A cell consisting of two dissimilar metals in contact with each other and with a common electrolyte (sometimes refers to two similar metals in contact with each other but with dissimilar electrolytes; differences can be small and more specifically defined as a concentration cell).

general corrosion Corrosion in a uniform manner.

gradation Sieve analysis of aggregates.

grade (1) Profile of the center of a roadway or the invert of a culvert or sewer. (2) The slope or ratio of rise or fall of the grade line to its length.

gradient See **grade**.

grain (1) A portion of a solid metal (usually a fraction of a centimeter [inch] in size) in which the atoms are arranged in an orderly pattern. The irregular junction of two adjacent grains is known as a grain boundary. (2) A unit of weight, 1/7 000th of a pound. (3) Used in connection with soil particles, as in a grain of sand.

granular The uniform size of grains of crystals in rock.

graphitization (graphitic corrosion) (1) Corrosion of gray cast iron in which the metallic constituents are converted to corrosion products, leaving the graphite flakes intact. (2) Used in a metallurgical sense to mean the decomposition of iron carbide to form iron and graphite.

groin A jetty built at an angle to the shore line to control the waterflow and currents or to protect a harbor or beach.

groundwater table (or level) Upper surface of the zone of saturation in permeable rock or soil. When the upper surface is confined by impermeable rock, the water table is absent.

grout (1) A fluid mixture of cement, water, and sometimes sand that can be poured or pumped easily. (2) Chemical mixtures recognized as stopping water infiltration through small holes and cracks.

grouting (1) The joining together of loose particles of soil in such a manner that the soil so joined becomes a solid mass impervious to water. (2) The process of flowing a cement and watergrout (without rock) into the annular space between a host pipe and a slipline pipe. Also see **pipe joint sealing**.

gunite or **reinforced shotcrete** A pipe-lining mixture of fine aggregate, cement, and water applied by air pressure using a cement ejector. Gunite is denser and has a higher ultimate compressive strength than cement mortar. Gunite is ideal for extremely deteriorated large sewers where persons and equipment can work without restriction.

head (static) (1) The height of water above any plane or point of reference. (2) The energy possessed by each unit of weight of a liquid, expressed as the vertical height through which a unit of weight would have to fall to release the average energy possessed. Standard unit of measure is the meter (foot). Relation between pressure expressed in kPa (psi or psf) of head is

$$\text{psf} = \frac{\text{lb/sq in.} \times 144}{\text{density in lb/cu ft}}$$

Head in feet for water at 68°F
1 lb/sq in. = 2.310 ft

headwall A wall of any material at the end of a culvert or drain to serve one or more of the following purposes: protect fill from scour or undermining; increase hydraulic efficiency; divert direction of flow; or serve as a retaining wall.

height of cover (HC) Distance from crown of a culvert or conduit to the finished road surface, ground surface, or the base of the rail.

high-density polyethylene (HDPE) A plastic resin made by the copolymerization of ethylene and a small amount of another hydrocarbon. The resulting base resin density, before additives or pigments, is greater than 0.941 g/mL.

holiday Any discontinuity or bare spot in a coated surface.

hoop stress The circumferential force per unit area in the pipe wall caused by internal pressure in kPa (psi).

hydraulic cleaning Techniques and methods used to clean sewer lines with water, such as water pumped in the form of a high-velocity spray and water flowing by gravity or head pressure. Devices include high-velocity jet cleaners, cleaning balls, and hinged-disc cleaners.

hydraulic gradient or **hydraulic grade line** An imaginary line through the points to which water would rise in a series of vertical tubes connected to the pipe. In an open channel, the water surface itself is the hydraulic grade line.

hydraulic radius or **hydraulic mean depth** The area of the water prism in the pipe or channel divided by the wetted perimeter. Thus, for a round conduit flowing full or half full, the hydraulic radius is $d/4$.

hydraulics That branch of science or engineering that treats water or other fluid in motion.

hydrocarbons, gaseous An organic compound made up of the elements of carbon and hydrogen that exists as a gas at ambient conditions (101.4 kPa, 23°C [14.7 psi, 73.4°F]).

hydrocarbons, liquid An organic compound made up of the elements of carbon and hydrogen that exists as a liquid at ambient conditions (101.4 kPa, 23°C [14.7 psi, 73.4°F]).

hydrogen blistering Subsurface voids produced in a metal by hydrogen absorption in (usually) low strength alloys with resulting surface bulges.

hydrogen induced cracking (HIC) A form of hydrogen blistering in which stepwise internal cracks are created that can affect the integrity of the metal.

hydrogen ion (pH) Refers to acidity or alkalinity of water or soil. An ion is a charged atom or group of atoms in solution or in a gas. Solutions contain equivalent numbers of positive and negative ions.

hydrogen stress cracking A cracking process that results from the presence of hydrogen in a metal in combination with tensile stress. It occurs most frequently with high-strength alloys.

hydrostatic design basis Hydrostatic design basis can be defined as the nominalized long-term strength or calculated hoop strength of the material at 100 000 hours obtained by long-term hydrostatic testing of pipe samples from which the probable safe life of the pipe at various stress levels (working pressures) and at various temperatures can be predicted.

ignition temperature Temperature at which the vapors emitted from a material will ignite either without exposure to a flame (self ignition) or when a flame is introduced (flash ignition).

impact Stress in a structure caused by the force of a vibrating, dropping, or moving load. This is generally a percentage of the live load.

impact strength The ability of a material to withstand shock loading.

impervious Impenetrable; completely resisting entrance of liquids.

inert material A material that is not very reactive, such as a noble metal or plastic.

infiltration The water entering a sewer system, including building sewers, from the ground through such means as defective pipes, pipe joints, connections, or manhole walls. Infiltration does not include and is distinguished from **inflow**.

infiltration and inflow (I/I) A combination of infiltration and inflow wastewater volumes in sewer lines, with no way to distinguish either of the basic sources, and with the same effect of usurping the capacities of sewer systems and facilities.

inflow The water discharged to a sewer system, including service connections, from such sources as roof leaders; cellar, yard, and area drains; foundation drains; cooling water discharges; drains from springs and swampy areas; manhole covers; cross-connections from storm sewers, combined sewers, catch basins; storm waters; surface runoff; street washwaters; or drainage. Inflow does not include and is distinguished from infiltration.

inhibitor (1) A chemical substance or combination of substances that, when present in the environment, prevents or reduces corrosion without significant reaction with the components of the environment. (2) A chemical additive that delays the chemical reaction in resin systems.

inspector The owner's on-site representative responsible for inspection and acceptance, approval, or rejection of work performed as set forth in specifications.

Inspector An authorized representative of the engineer assigned to observe the work performed and materials furnished by the contractor or such other person as may be appointed by the owner as his representative. The contractor shall be notified in writing of the identity of the representative. Also referred to as construction observer, resident inspector, construction inspector, or project representative.

interaction The division of load carrying between pipe and backfill and the relationship of one to the other.

intercepting drain A ditch or trench filled with a pervious filter material around a subdrainage pipe.

interceptor sewer A sewer that receives the flow from collector sewers and conveys the wastewater to treatment facilities.

intergranular stress corrosion cracking (IGSCC) Stress corrosion cracking in which the cracking occurs along grain boundaries.

internal corrosion Corrosion that occurs inside a pipe because of the physical, chemical, or biological interactions between the pipe and the water as opposed to forces acting outside the pipe, such as soil, weather, or stress conditions.

internal erosion Abrasion and corrosion on the inside diameter of the pipe or tubing caused by the fluid being transported.

internal pipe inspection The television inspection of a sewer line section. A closed-circuit television camera is moved through the line at a slow rate and a continuous picture is transmitted to an aboveground monitor. Also see **physical pipe inspection**.

intrados The inner curve of an architectural arch.

inversion The process of turning a fabric tube inside out with water or air pressure as at the installation of a cured-in-place pipe.

invert That part of a pipe or sewer below the spring line, generally the lowest point of the internal cross section.

invert level (elevation) The level (elevation) of the lowest portion of a liquid-carrying conduit, such as a sewer, which determines the hydraulic gradient available for moving the contained liquid.

ion An electrically charged atom (Na^+, Al^{3+}, Cl^-, S_2^-) or group of atoms known as "radicals" (NH_4^+, SO_4^{2-}, PO_4^{3-}).

ionization Dissociation of ions in an aqueous solution (for example, $H_2CO_3 \rightarrow H^+ + HCO_3^-$ or $H_2O \rightarrow H^+ + OH^-$).

jacking (for conduits) A method of providing an opening for drainage or other purposes underground, by cutting an opening ahead of the pipe and forcing the pipe into the opening by means of horizontal jacks.

joints The means of connecting sectional lengths of sewer pipe into a continuous sewer line using various types of jointing materials. The number of joints depends on the lengths of the pipe sections used in the specific sewer construction work.

kip A stress unit equal to 4.448 kN (1 000 lb).

lateral Any pipe connected to a sewer.

linear foot The unit of measurement relating to the length of a sewer line.

long-term strength The hoop stress in the wall of the thermoplastic pipe in the circumferential direction that when applied continuously will cause failure of the pipe at 100 000 hours.

major blockage A blockage (structural defect, collapse, protruding service connection, debris) that prohibits manhole-to-manhole cleaning, television inspection, pipe flow, or rehabilitation procedures.

manhole section The length of sewer pipe connecting two manholes.

Manning's formula (1) An equation for the value of coefficient (C) in the Chezy formula, the factors of which are the hydraulic radius (R_H) and a coefficient of roughness ($\in$). (2) An equation itself used to calculate flows in gravity channels and conduits.

maximum allowable operating pressure The highest working pressure expected and designed for during the service life of the main.

mechanical cleaning Methods used to clean sewer lines of debris mechanically with devices such as rodding machines, bucket machines, and winch-pulled brushes.

median barrier A double-faced guardrail in the median or island dividing two adjacent roadways.

melt flow A measure of a molten material's fluidity.

melt flow rate The quantity of thermoplastic material, in grams, that flows through an orifice during a 10-minute time span under conditions as specified by ASTM D-1238.

melt index The melt flow of a thermoplastic material as determined under Condition E of ASTM D-1238. Condition C or high load melt index produces a higher melt flow for a given material as a result of the greater force applied to the sample during the test.

melt viscosity The resistance of a molten material to flow.

modifications (1) A written amendment of the contract documents signed by both parties. (2) A change order. (3) A written clarification or interpretation issued by the engineer according to the terms of the contract. (4) A written order for a minor change or alteration in the work issued by the engineer pursuant to the terms of the contract. A modification may only be issued after execution of the agreement.

modulus of elasticity (E) The stress required to produce unit strain, which may be a change of length (Young's modulus), a twist or shear (modulus), or a change of volume (bulk modulus), expressed in dynes per square centimeter.

molecular weight distribution The ratio of the weight average molecular weight to the number average molecular weight. This gives a preliminary indication of the range of molecular weights.

moment, bending The moment that produces bending in a beam or other structure. It is measured by the algebraic sum of the products of all the forces multiplied by their respective lever arms.

moment of inertia Function of some property of a body or figure (such as weight, mass, volume, area, length, or position) equal to the summation of the products of the elementary portions by the squares of their distances from a given axis.

neutral axis An axis of no stress.

nominalize To classify a value into an established range or category.

nonuniform corrosion Corrosion that attacks small, localized areas of the pipe, usually resulting in less metal loss than uniform corrosion but causing more rapid failure of the pipe because of pits and holes.

notch sensitivity The extent to which an inclination to fracture is increased by a notch, crack, scratch, or sudden change in section.

notice of award The written notice by owner to the apparent successful bidder stating that, upon compliance with the conditions to be fulfilled by him within the time specified, the owner will execute and deliver the agreement to him.

notice to proceed A written notice given by the owner to the contractor with a copy to the engineer fixing the date on which the contract time will commence to run and on which the contractor will start to perform his obligations under the contract documents.

outfall or **outlet** In hydraulics, the discharge end of drains and sewers.

overflow (1) The excess water that flows over the ordinary limits of a sewer, manhole, or containment structure. (2) An outlet, pipe, or receptacle for excess water.

owner A public body of authority, corporation, association, partnership, or individual for whom the work is to be performed.

oxidation Loss of electrons, as when a metal goes from the metallic state to the corroded state.

parapet (1) Wall or rampart, breast high. (2) The wall on top of an **abutment** extending from the bridge seat to the underside of the bridge floor and designed to hold the backfill.

Pascal's law Pressure exerted at any point upon a confined liquid is transmitted undiminished in all directions.

pavement, invert Lower segment of a corrugated metal pipe provided with a smooth bituminous material that completely fills the corrugations, intended to give resistance to scour and erosion and improve flow.

perched water table In hydrology, the upper surface of a body of free groundwater in a zone of saturation, separated by unsaturated material from an underlying body of groundwater in a differing zone of saturation.

periphery Circumference or perimeter of a circle, ellipse, pipe-arch, or other closed curvilinear figure.

permeability Penetrability.

pH A measure of the acidity or alkalinity of a solution. A value of 7 is neutral; lower numbers are acidic, higher numbers are alkaline.

physical pipe inspection (man entry) The crawling or walking through manually accessible pipe lines. Manual inspection is only undertaken when field conditions permit this to be done safely. Precautions are necessary.

pile, bearing A member driven or jetted into the ground and deriving its support from the underlying strata or by the friction of the ground on its surface.

pipe joint sealing A method of sealing leaking or defective pipe joints that permit infiltration of groundwater to sewers by means of injecting chemical grout to or through the joints from within the pipe.

pipeline reconstruction The *in situ* repair of an existing pipeline that has suffered loss of pressure integrity or has been structurally damaged. The liner becomes the principal pressure containment or structural element of the *in situ* composite pipe structure.

pipeline rehabilitation The *in situ* repair of a corroded or abraded pipeline by insert renewal of a liner that rehabilitates the bore of the pipeline and improves flow efficiency or hydraulics but does not contribute significantly to increased pressure capability or increased structural strength.

pitting Highly localized corrosion resulting in deep penetration at only a few spots.

pitting factor The depth of the deepest pit divided by the "average penetration" as calculated from weight loss.

plant piping Installation procedure that digs a trench and lays the pipe in one step.

Plastic Pipe Institute (PPI) A division of the Society for the Plastics Industry, Inc.

plate A flat-rolled iron or steel product.

plough-in piping Installation procedure that splits the earth and pulls the pipe into position.

polyester Resin formed by condensation of polybasic and monobasic acids with polyhydric alcohols.

polyethylene A ductile, durable, virtually inert thermoplastic composed by polymers of ethylene. It is normally a translucent, tough solid. In pipe grade resins, ethylene-hexene copolymers are usually specified with carbon black pigment for weatherability.

ponding (1) Jetting or the use of water to hasten the settlement of an embankment, using the judgment of a soils engineer. (2) In hydraulics, ponding refers to water backed up in a channel or ditch as the result of a culvert of inadequate capacity or design to permit the water to flow unrestricted.

precipitation Process by which water in liquid or solid state (rain, sleet, snow) is discharged out of the atmosphere to a land or water surface.

pressure rating Estimated maximum internal pressure that allows a high degree of certainty that failure of the pipe will not occur.

primary properties The properties used to classify polyethylene materials.

profile Anchor pattern on a surface produced by abrasive blasting or acid treatment.

project The entire construction to be performed as provided in the contract documents.

pull-in piping Also referred to as insert renewal; installation procedure whereby pipe is pulled inside old mains and service lines to provide the new main or service line.

radian (1) An arc of a circle equal in length to the radius. (2) The angle at the center measured by the arc.

radius of gyration The distance from the reference where all of the area can be considered concentrated that still produces the same moment of inertia. Numerically, it is equal to the square root of the moment of inertia divided by the area.

rainfall Precipitation in the form of water or snow.

reduction Gain of electrons, as when copper is electroplated on steel from a copper sulfate solution (opposite of oxidation).

regression analysis An evaluation of the long-term hoop stress data. A linear curve is calculated using the least-squares method to fit the logarithm of hoop stress versus the logarithm of the resulting hours-to-failure.

regulator A device for controlling the quantity of wastewater and stormwater admitted from a combined sewer collector line to an interceptor sewer, pumping station, or treatment facility, thereby determining the amount and quality of the flows discharged through an overflow device to receiving waters or other points of disposal.

rehabilitation All aspects of upgrading the performance of existing sewer systems. Structural rehabilitation includes repair, renovation, and renewal. Hydraulic rehabilitation covers replacement, reinforcement, flow reduction or attenuation, and (occasionally) renovation.

reinforced shotcrete See **gunite**.

reinforcement The provision of an additional sewer that, with an existing sewer, increases overall flow capacity.

renewal Construction of a new sewer, on or off the line of an existing sewer. The basic function and capacity of the new sewer are similar to those of the existing one.

renovation Methods by which the performance of a length of sewer is improved by incorporating the original sewer fabric but excluding maintenance operations such as isolated local repairs and root or silt removal.

repair Rectification of damage to the structural fabric of the sewer and the reconstruction of short lengths but not the reconstruction of the whole pipeline.

replacement Construction of a new sewer, on or off the line of an existing sewer. The function of the new sewer will incorporate that of the existing one but may also include improvement or development.

resin impregnation (wet-out) A process used in cured-in-place pipe installation where a plastic-coated fabric tube is uniformly saturated with a liquid thermosetting resin while air is removed from the coated tube by means of vacuum suction.

resins An organic polymer, solid, or liquid; usually thermoplastic or thermosetting.

retaining wall A wall for sustaining the pressure of earth or filling deposited behind it.

revetment A wall or a facing of wood, willow mattresses, steel units, stone, or concrete placed on stream banks to prevent erosion.

Reynolds number A dimensionless quantity named after Osbourne Reynolds, who first made known the difference between laminar and turbulent flow. The practical value of the Reynolds number is that it indicates the degree of turbulence in a flowing liquid. It depends on the hydraulic radius of the conduit, the viscosity of the water and the velocity of flow. For a conduit of a given size, the velocity is generally the major variable and the Reynolds number will increase as the velocity of flow increases.

right bank That bank of a stream that is on the right when one looks downstream.

ring compression The principal stress in a confined thin circular ring subjected to external pressure.

riprap Rough stone of various sizes placed compactly or irregularly to prevent scour by water or debris.

roadway (1) Portion of the highway included between the outside lines of gutters or side ditches, including all slopes, ditches, channels, and appurtenances necessary for proper drainage, protection, and use. (2) That part of the railway right-of-way prepared to receive the track. During construction, the roadway often is referred to as the grade.

roof leader A drain or pipe that conducts stormwater from the roof of a structure downward and to a sewer for removal from the property or onto the ground for runoff or seepage disposal.

roughness coefficient A factor in the Kutter, Manning, and other flow formulae representing the effect of channel (or conduit) roughness on energy losses in the flowing water.

runoff (1) That part of precipitation carried off from the area on which it falls. Also, the rate of surface discharge of the above. (2) That part of precipitation reaching a stream, drain or sewer. Ratio of runoff to precipitation is a coefficient expressed decimally.

samples Physical examples that illustrate materials, equipment, or workmanship and establish standards by which the work will be judged.

sanitary sewer A sewer intended to carry only sanitary or sanitary and industrial wastewater from residences, commercial buildings, industrial parks, and institutions.

scaling High-temperature corrosion resulting in formation of thick corrosion product layers. Deposition of insoluble materials on surfaces, usually inside water boilers or heat exchanger tubes.

secondary stress Forces acting on the pipe in addition to the internal pressure, such as forces imposed by soil loading and dynamic soil conditions.

sectional properties End area per unit of width, moment of inertia, section modulus, and radius of gyration.

section modulus The moment of inertia of the area of a section of a member divided by the distance from the center of gravity to the outermost fiber.

seepage Water escaping through or emerging from the ground along some rather extensive line or surface, as contrasted with a spring, the water of which emerges from a single spot.

serviceability of the piping system Continued service life with a high degree of confidence that a failure will not occur during its long-term service.

sewer building The conduit that connects building wastewater sources to the public or street sewer, including lines serving homes, public buildings, commercial establishments, and industry structures. In a specification, the building sewer is referred to in two sections: (1) the section between the building line and the property line, frequently specified and supervised by plumbing or housing officials, and (2) the section between the property line and the street sewer and the included connection, frequently specified and supervised by sewer, public works, or engineering officials. Also referred to as house sewer, building connection, service connection.

sewer cleaning The use of mechanical or hydraulic equipment to dislodge, transport, and remove debris from sewer lines.

sewer interceptor A sewer that receives the flow from collector sewers and conveys the wastewater to treatment facilities.

sewer pipe A length of conduit, manufactured from various materials and in various lengths, that when joined together can be used to transport wastewater from the points of origin to a treatment facility. Types of pipe include acrylonitrile-butadiene-styrene (ABS); asbestos-cement (AC); brick pipe (BP); concrete pipe (CP); ductile iron pipe (DIP); polyethylene (PE); polyvinyl chloride (PVC); vitrified clay (VC); and reinforced plastic mortar (RPM).

shaft A pit or well sunk from the ground surface into a tunnel for the purpose of furnishing ventilation or access to the tunnel.

sheeting A wall of metal plates or wood planking to keep out water or soft or flowing materials.

shop drawings All drawings, diagrams, illustration, brochures, schedules, and other data that are prepared by the contractor, a subcontractor, manufacturer, supplier, or distributor that illustrate the equipment, material or some portion of the work as required by the contract documents.

siphon (inverted) A conduit or culvert with a U- or V-shaped grade line to permit it to pass under an intersecting roadway, stream or other obstruction.

site Any location where work has been or will be done.

site access An adequately clear area of a size sufficient to accommodate personnel and equipment required at the location where work is to be performed, including roadway or surface sufficiently unobstructed to permit conveyance of vehicles from the nearest paved roadway to the work location.

skew (or skew angle) The acute angle formed by the intersection of the line normal to the centerline of the road improvement with the centerline of a culvert or other structure.

slide Movement of a part of the earth under force of gravity.

smooth radius bend A contoured sweep or bend with no sharp or angular sections.

social costs Costs incurred by society as a result of sewerage works and for which authorities have no direct responsibility. These include unclaimed business losses from road closures, and the cost of extended journey times because of traffic diversions.

softening temperature There are many ways to measure the softening temperature of a plastic. The commonly reported Vicat Softening Temperature method is to measure the temperature at which penetration of a blunt needle through a given sample occurs under conditions specified in ASTM D-1525.

solar radiation The emission of light from the sun, including short ultraviolet wavelengths, visible light, and long infrared wavelengths.

solubility The amount of one substance that will dissolve in another to produce a saturated solution.

spalling The spontaneous chipping, fragmentation, or separation of a surface or surface coating.

span Horizontal distance between supports or maximum inside distance between the sidewall of culverts.

special conditions When included as a part of the contract documents. Special conditions refer only to the work under the contract.

specifications Those portions of the contract documents consisting of written technical descriptions of materials, equipment, construction systems, standards and workmanship as applied to the work.

specific gravity The density of a material divided by the density of water, usually at 4°C. Because the density of water is nearly 1 g/cu cm, density in g/cu cm and specific gravity are approximately equal.

spelter Zinc or galvanized coating on steel products.

spillway (1) A low-level passage serving a dam or reservoir through which surplus water may be discharged; usually an open ditch around the end of a dam or a gateway or a pipe in a dam. (2) An outlet pipe, flume or channel serving to discharge water from a ditch, ditch check, gutter, or embankment protector.

springing line (1) Line of intersection between the **intrados** and the supports of an arch. (2) The maximum horizontal dimension of a culvert or conduit.

spring line The horizontal midpoint of a sewer pipe.

spun lining A bituminous or cement mortar lining in a pipe made smooth or uniform by spinning the pipe around its axis.

stabilizer An ingredient used in the formulation of some plastics to assist in maintaining the physical and chemical properties of the compounded materials at their initial values throughout the processing and service life of the material.

Standard dimension ratio (SDR) The ratio of the pipe diameter to wall thickness.

storm sewer A sewer intended to carry only stormwater, surface runoff, street washwater, and drainage.

strain The movement of a material from a given stress on the material. Usually expressed as inches per inch of movement (centimeters per centimeter of movement).

strain corrosion (SC) Microcracking or crazing of a plastic material permitting a corrosive agent to attack various impurities in the pipe wall.

strain rate The rate of lineal change per unit length.

street access Areas normally used for public vehicular traffic (including roads, streets, or rights-of-way) to which safe access is required for performance of work.

stress The load applied per unit area of material. Often expressed as kPa (psi).

stress corrosion cracking (SCC) Cracking of a metal produced by the combined action of corrosion and tensile stress (residual or applied).

stress life curves Graphic representations showing the extrapolation to 100 000 hours of applied hoop stress versus hours to failure data on logarithmic scales per ASTM D-2837 and D-2992.

stress relief The decrease in imposed stresses at a constant strain. In plastic materials it occurs as a property of the material with time.

subcontractor An individual, firm, or corporation having a direct contract with the contractor or any other subcontractor for the performance of a part of the work at the site.

subdrain A pervious backfilled trench containing a pipe with perforations or open joints for the purpose of intercepting groundwater or seepage.

subgrade The surface of a portion of the road bed on which paving, railroad track ballast, or any other structure is placed.

substantial completion The stage in construction when a project can be used for its intended purposes. At substantial completion, minor items and items that are seasonally restricted need not be completed, but the items that affect operational integrity and function of the facility must be capable of continuous use.

sulfide stress cracking (SSC) Brittle failure by cracking under the combined action of tensile stress and corrosion in the presence of water and hydrogen sulfide.

supplementary general conditions Federal conditions in effect at the time of submission of the bid.

supplier Any person or organization who supplies materials or equipment for the work, including that fabricated to a special design, but who does not perform labor at the site.

surcharge When the sewer flow exceeds the hydraulic carrying capacity of the sewer line.

surcharge condition When the sewer flow depth equals or exceeds the diameter of the discharging sewer line.

surety The corporate body that is bound with the contractor and that engages to be responsible for the contractors and their acceptable performance of the work.

surface hardness A measure of the net increase in depth as an indented load is increased from a minor load to a major load and then returned to a minor load. This is used as an indication of relative hardness among like materials.

swale (DIP, SAG) A significant deviation in pipe grade, such as to cause entrapment of solids, semisolids, and liquids, thereby impeding the accuracy or effectiveness of flow measurements, cleaning, and internal inspection.

tailwater The water just downstream from a structure.

tear strength An indication of the relative strength required to force propagation of an induced imperfection.

tensile strength The pulling stress, in kPa (psi), required to elongate a given specimen to the breaking point.

tensile strength at yield The measured tensile stress required to initiate permanent deformation in a sample under the conditions described in ASTM D-638.

test medium The fluid or gas inside the main being tested.

thermal conductivity The ability of a material to conduct heat; a physical constant for the quantity of heat that passes through a unit cube of material in a unit of time when the temperature difference is 0.56°C (1°F).

thermal expansion contraction The fractional change in length of a material subjected to a unit change in temperature.

thermal stabilizers Compounds added to the plastic resins when compounded that prevent degradation of properties due to elevated temperatures.

thermoplastic A material, such as polyvinyl chloride, that will soften when heated and harden when cooled.

thermoset A material, such as epoxy, that will undergo or has undergone a chemical reaction by the action of heat, a chemical catalyst, or ultraviolet light, leading to an infusible state.

threading The process of installing a slightly smaller pipe or arch within a failing drainage structure.

toe drain A subdrain installed near the downstream toe of a dam or levee to intercept seepage.

tuberculation Localized corrosion at scattered locations resulting in knoblike mounds.

ultraviolet absorbers (stabilizers) Compounds that when mixed with thermoplastic resin selectively absorb ultraviolet rays, protecting the resins from ultraviolet attack.

underdrain See **subdrain**.

uniform corrosion Corrosion that results in an equal amount of material loss over an entire pipe surface.

velocity head For water moving at a given velocity, the equivalent head through which it would have to fall by gravity to acquire the same velocity.

voids A term generally applied to paints or coatings to describe holidays, holes, and skips in the film. Also used to describe shrinkage in castings or welds.

wale Guide or brace of steel or timber, used in trenches and other construction.

watershed Region or area contributing to the supply of a stream, or lake, drainage area, drainage basin, or catchment area.

water table The upper limit of the portion of ground wholly saturated with water.

weatherability The properties of a plastic material that allows it to withstand natural weathering (hot and cold temperatures, wind, rain, and ultraviolet rays).

wetted perimeter The length of the perimeter in contact with the water. For a circular pipe of inside diameter d, flowing full, the wetted perimeter is the circumference, πd. The same pipe flowing half full would have a wetted perimeter of $d/2$.

work Any and all obligations, duties, and responsibilities necessary to the successful completion of the project assigned to or undertaken by the contractor under the contract documents, including all labor, materials, equipment, other incidentals, and their furnishing.

written notice Written demands, instructions, claims, approvals, and disapprovals required to obtain compliance with contract requirements. (2) Written notice is served if delivered in person to the individual or to a member of the corporation for whom it is intended, or to an authorized representative of such individual or corporation, or if delivered at or sent by registered mail to the last business address known. Unless otherwise stated in writing, any notice to or demand upon the owner under this contract shall be delivered to the owner through the **engineer**.

Index

A

Acrylamide gel, 181, 183
Acrylamide grout, 208
Acrylate gel, 181
Acrylate grout, 209
Acrylic grout, 209
Air flotation, CSOs, 97
Annular grouting, 150, 216
Atmospheric hazards, 51
   explosive gases, 51
   oxygen deficiency, 51
   toxic conditions, 51
Auger boring, 173
Auger systems, 169

B

Base flow, 40
Brick sewers, 28
Bubbler recorders, 122
Bucket test, 121
Building inspection, 53, 54
Bypasses, flow monitoring, 105

C

Calibration, pumping station, 118
Camera-packer method, 196
Capacitance recorders, 122
Cast-in-place concrete linings, 163
Catch basins, 225
Cement grout, 184
Cement-lined ductile iron pipe, 156
Chemical grouting, 207
   external pipeline, 183
   internal, 178
   manhole, 189
   service connections, 196
Chemical tracers, 117
Chemical treatment, CSOs, 97
Cipolletti weir, 113
Clay sewers, 29
Cleaning, 175
Coating compounds, 190
Coatings, 162, 207
   manhole, 189
Combined sewers, 6
Compaction grouting, 185
Compound weir, 113
Concrete sewers, 29
Confined space entry, 148
Construction, preparation for, 148
Continuous pipe sliplining, 152, 200
Contracted weir, 112
Contracts, construction, 217
Conventional replacement, 173
Corrosion protection, manhole, 193
Costs
   grouting, 182
   rehabilitation, 147
Critical sewers, 13
   categories, 13
   definitions, 14
CSO abatement
   data requirements, 89
   flow monitoring, 90
   investigation, 89
   issues in, 87
   modeling, 91
   planning, 87

problems, 86
regulations, 86, 88
sampling, 91
treatment strategies, 96
upgrading/treatment options, 94
CSO effects, 12
environmental, 13
flooding, 13
Cured-in-place pipe, 203
pipeline, 156
service connections, 196
Current meters, 123
Cutting, 176

D

Data correlation, flow monitoring, 106
Data needs, flow monitoring, 100
Data recording, 53
Deformed pipe, 158, 204
Depth meters, 126
Depth recorders, 121
Depth-velocity meters, 126
Differential isolation, 62
Digging, 176
Directional drilling, 172
Discharge curves, 110
Disinfection, CSOs, 97
Documentation, contract, 151
Doppler meters, 123
Dragging, 176
Ductile iron pipe, 156, 203
Dye-dilution technique, 117
Dye tracers, 117
Dye-water testing, 55
isolation, 60
logging results, 58
manholes, 56
night flow isolation, 58
private property, 56
safety measures, 57
storm sewer sections, 56
stream sections, 56
suspected sources, 57

E

Electromagnetic meters, 123
Electronic recorders, 122
Environmental effects, 13
Evaluation assessment
monitoring, 33
rehabilitation, 34
replacement, 35
stabilization, 34
Evaluation techniques
leak quantification, 66
photographs, 66
pipeline cleaning, 65
results, 64
television inspection, 65
Explosive gases, 51
External grouting, 183

F

Fiber glass pipe, 202
Fiber-glass-reinforced cement linings, 161, 206
Fiber-glass-reinforced plastic linings, 162, 206
Fiber-glass-reinforced plastic pipe, 155
Field investigations, CSOs, 90
Filtration, CSOs, 97
Float recorders, 122
Flooding, 13
Flow attenuation, CSOs, 95
Flow components, 40
Flow data
evaluation, 44
normalization, 218
storm inflow comparison, 219
Flow diversion, 149
Flow measurement, 109
continuous, 126
manual, 124
Flow metering equipment, maintenance, 124

Flow metering, preinstallation, 42
Flow meters
   automatic, 121
   maintenance, 43
Flow monitoring, 90
   data analysis, 129
   programs, 103
   quality assurance, 130, 220
Flow volumes, annual, 41
Fluid jet cutting, 172
Flumes, 114
Fluorometric methods, 64
Foam, 208
Foundation drain connections, 225

G

Gel, 181, 207
Groundwater
   gauging, 45
   level measurement, 128
Groundwater levels, infiltration, 72
Groundwater migration, 59
Grouting, 178
   annular, 150
Grout sealing rings, 179
Gunite, 163

H

H-flume, 117
High-density polyethylene pipe, 200, 201
Hydraulic capacity, 145
Hydraulic performance, 16
Hydraulic problems, 9
   infiltration, 9
   inflow, 12
Hydraulics, assessing, 21

I

I/I condition, assessing, 20
I/I control
   measuring effectiveness, 217
   objectives, 214
   rehabilitation effectiveness, 220
I/I evaluation
   dye-water testing, 55
   flow components, 40
   precipitation measurement, 44
   smoke testing, 46
   techniques, 40
I/I problems, 40
Impact moling, 172
Impact ramming, 172
Increasing system capacity, CSOs, 95
Infiltration, 9
Infiltration evaluation, 68
   location of, 71
   small subsystems, 70
   water use, 68
Infiltration location, 71
   cost effectiveness, 72
   groundwater levels, 72
   interviews, 72
   visual inspection, 72
Infiltration simulation, 218
Inflow, 12
Inflow control, quality assurance, 224
Inflow evaluation, 73
   cost effectiveness, 78
   methodology, 74
   rainfall-induced, 77
   techniques, 74
Inflow simulation, 219
Inspections, 49
   building, 53, 54
   construction, 217
   data recording, 53

safety measures, 50
television, 82
visual, 81
Instantaneous monitoring, 108
Internal grouting, 178
Inversion lining, quality assurance, 223
Investigations
CSOs, 90
cursory, 23
detailed, 18
planning, 17
Isolation
differential, 62
fluorometric methods, 64
plugging, 60
portable weirs, 62
velocity-area method, 63
Isolation and measurement, 124

J

Jet rodding, 176

L

Leak quantification, 66
Linings
manhole, 191
pipeline, 152
quality assurance, 222
segmental, 161, 206
service connections, 196, 197
structural design, 145
Loadings, structural, 145
Long-term monitoring, 109

M

Maintenance
continuing, 226
preventive, 227

Manholes
inflow, 226
inspections, 49
rehabilitation, 187
Manning's equation, 110, 126
Mapping, 100
CSOs, 90
I/I, 80
Materials
abrasion resistance, 145
chemical resistance, 144
durability, 141
maintenance, 207
properties, 144
quality control, 215
replacement, 207
Mechanical sealing, 185
Metering location, calibration, 127
Meter maintenance, 43
Microfine cement grout, 185
Microtunneling, 169
Modeling
CSOs, 91
hydraulic function, 25
hydrologic function, 23
justification for, 26
predictive, 23
time scale, 25
Monitoring, 33
Multipass isolation, 60

N

Night flow isolation, 58
groundwater migration, 59
multipass, 60
single-pass, 60
Nozzle meters, 123
NU-pipe, 158, 204

O

Orifice meters, 123

Outlet and junction measurements, 124
Overflows, flow monitoring, 105
Oxygen deficiency, 51

P

Palmer–Bowlus flume, 116
Parshall flume, 115
Patching compounds, 190
Peak rates, 41
Performance criteria, 15
Permanent monitoring, 109
Photographs, 66
Physical condition evaluation, 79
    mapping, 80
    O & M problems, 80
    structural integrity, 79
    television inspection, 82
    visual inspection, 81
Pipe bursting, 165
Pipe condition
    Condition I, 35
    Condition II, 35
    Condition III, 35
Pipe laying, 150
Pipeline cleaning, 65
Pipeline conditions, 27
    brick sewers, 28
    clay sewers, 29
    concrete sewers, 29
    corrosion, 28
    qualification, 28
    quantification, 28
    structural, 28
Pipeline inspections, 49
Plugging compounds, 190
Plugging isolation, 60
Pointing, 178
Point repair, 34, 185
Pollutant loads, CSOs, 94
Polybutylene pipe, 200
Polyethylene-lined ductile iron pipe, 156
Polyethylene linings, 162, 206
Polyethylene pipe, 154

Polyurethane foam, 181
Polyvinyl chloride
    linings, 162, 206
    pipe, 155, 201
Pomeroy equation, 127
Portable weirs, 62
Portland cement grout, 184
Precipitation measurement, 44
Predictive modeling, 23
Pressure grouting, 179
Pressure sensors, 122
Probe injection grouting, 180
Probe recorders, 121
Problems
    hydraulic, 9
    structural, 8
Profile wall polyethylene pipe, 154
Pump full method, 196
Pumping station calibration, 118
    discharge volume, 119
    pump curve, 120
    velocity meter, 119
    wet well drawdown, 118

Q

Quality assurance, construction-phase, 215
Quality control, materials, 215

R

Radioactive tracers, 117
Rainfall, measurement, 129
Random monitoring, 108
Real-time control, CSOs, 96
Records, 101
Rectangular weir, 112
Reinforced plastic mortar linings, 162, 206
Reinforced plastic mortar pipe, 156
Reinforced shotcrete, 163
Reinforced thermo setting resin pipe, 156

Replacement, 35
Reports, annual, 228
Rodding, 176
Rolldown pipe, 159, 204
Roof drains, 225
Root control, 176, 207

S

Safety
   construction, 148, 216
   flow monitoring, 107
   inspections, 50
Safety measures
   animals, 52
   atmospheric hazards, 51
   chemicals, 52
   drowning, 52
   dye-water testing, 57
   infections, 52
   physical injury, 51
Sampling, 91
Sanitary flow simulation, 219
Sanitary sewers, 6
Screening, CSOs, 96
Sealing, quality assurance, 223
Sedimentation structures, CSOs, 97
Segmental linings, 161, 206
Service connection rehabilitation, 194
Service laterals, quality assurance, 224
Sewer cleaning, 149
Sewer preparation, 216
Sewer rehabilitation, objectives, 213
Sewer replacement, quality assurance, 221
Sewer sausage method, 196
Sewer stabilization, 34
Sewer system control, CSOs, 95
Sewer system evaluation, groundwater gauging, 45
Sewers, categories of, 13
Short pipe sliplining, 153, 201
Short-term monitoring, 108
Shotcrete, 163

Simulation
   infiltration, 218
   inflow, 219
   sanitary flow, 219
Single-pass isolation, 60
Site selection, flow monitoring, 104
Site supervision, 150
Sliplining, 152, 200
   quality assurance, 222
Slurry systems, 170
Smoke testing, 46
Soil stabilization, 34
Solid-wall polyethylene pipe, 154
Source control, CSOs, 94
Specifications, installation, 216
Spiral Rib polyethylene pipe, 155
Spiral-wound pipe, 161, 205
Spot repair, 34, 185
Stabilization methods, 34
Stage discharge curves, 110
Stage measurement, 121
Steel linings, 162, 207
Steel pipe, 156, 204
Storm sewers, 6
Structural condition, assessing, 20
Structural evaluation
   evaluation assessment, 32
   methods of, 27
   program plan, 32
   rehabilitation assessment, 30
Structural integrity
   evaluation, 79
   quality assurance, 221
Structural investigation, 215
Structural performance, 15
Structural problems, 8
   collapse mechanisms, 9
Sump pumps, 225
Suppressed weir, 112
Swagedown pipe, 159
System condition, assessing, 19

T

Television inspection, 65, 82
Toxic conditions, 51

Trapezoidal flume, 117
Trapezoidal weir, 113
Trench safety, 148
Trenchless replacement, 165

U

U-liner pipe, 158, 204
Ultrasonic recorders, 122
Urethane foam, 210
Urethane gel, 181
Urethane grout, 209
Usage plan
   developing, 22
   implementing, 22

V

Velocity measurement, 110
Velocity meters, 123
Velocity probes, 124
Velocity-area methods, 63
Venturi meters, 123
Visual inspections, 49, 72, 81
V-notch weir, 111

W

Water use evaluation, 68
   maximum daily flow, 69
   maximum-minimum daily flow, 69
   nighttime domestic flow, 70
   small subsystems, 70
Weirs, 111
Welded steel linings, 207
Winching, 176

Index